PATTERN RECOGNITION

ELECTRICAL ENGINEERING AND ELECTRONICS

A Series of Reference Books and Textbooks

Editors

Marlin O. Thurston
Department of Electrical
Engineering
The Ohio State University
Columbus, Ohio

William Middendorf
Department of Electrical
and Computer Engineering
University of Cincinnati
Cincinnati, Ohio

Other Volumes in Preparation

PATTERN RECOGNITION

Applications to
Large Data-Set Problems

Sing-Tze Bow
Department of Electrical Engineering
The Pennsylvania State University
University Park, Pennsylvania

MARCEL DEKKER, INC.   New York and Basel

Library of Congress Cataloging in Publication Data

Bow, Sing-Tze, [Date]
    Pattern recognition.

    (Electrical engineering and electronics ; 23)
    Bibliography: p.
    Includes index.
    1. Pattern recognition systems. I. Title.
II. Series.
TK7882.P3B69 1984   001.64'42   84-11420
ISBN 0-8247-7176-1

MARCEL DEKKER, INC.
270 Madison Avenue, New York, New York 10016

Current printing (last digit):
10 9 8 7 6 5 4 3 2 1

PRINTED IN THE UNITED STATES OF AMERICA

Preface

The purpose of writing this book is twofold: (1) to provide readers with the basic principles involved with the approaches currently employed in pattern recognition; and (2) to demonstrate use of the theories with relevant practical problems so that the theories may be better understood.

The materials collected in this book are grouped into two main parts. Part I emphasizes the principles of pattern recognition, and Part II deals with the preprocessing of data for pattern recognition. There are five chapters in Part I. Chapter 1 presents the fundamental concept of pattern recognition and its system configuration. Selected applications, including weather forecasting, handprinted character recognition, speech recognition, medical analysis, and satellite and aerial-photo interpretation, are discussed briefly. Also in Chapter 1, the two principle approaches used in pattern recognition, the decision theoretic and syntactic approaches, are described and compared. The remaining chapters in Part I focus primarily on the decision theoretic approach. Because of space limitations, the syntactic approach is left for another publication. Chapters 2 and 3 discuss some principles involved in nonparametric decision theoretic classification and the training of the discriminant functions used in these classifications. Chapter 4 introduces the principles of statistical decision theory in classification problems. In recent years, a great many advances have been made in the field of clustering, but because of space limitations and the need to be systematic, the material in Chapter 5 is selected and organized to make readers aware of current trends and the main thrust in attempts to solve pattern recognition problems, so that readers will have no difficulty following the current literature after they read this book.

In Part II emphasis is on the preprocessing of original data for accurate and correct pattern recognition. Appropriate preprocessing of original data has a considerable effect on proper selection of a method for pattern recognition. Chapter 6 discusses dimensionality reduction and feature selection, which are necessary measures in making machine recognition feasible. In that chapter attention is given to the optimum number of features and their ordering, to canonical analysis and its application to large data-set problems, and to the nonparametric feature selection method, which is applicable to pattern recognition problems based on mixed features. Chapters 7 and 8 are devoted primarily to the methodology employed in preprocessing a large data-set problem. More concretely, complex problems such as scenic images are used for illustration. Processing in both the spatial and transform domains is considered in detail.

A set of seven 512 × 512 256-gray-level images are included in Appendix A. These images can be used as large data sets to illustrate many of the pattern recognition and data preprocessing concepts developed in the text. They can be used in their original form and can be altered to generate a variety of input data sets.

This book is the outgrowth of two graduate courses developed for the Department of Electrical Engineering of The Pennsylvania State University: "Principles of Pattern Recognition" and "Digital Image Processing." This material has been rewritten to suit both graduate students and senior undergraduates with high grade-point averages. The book can be used as a one-semester course on pattern recognition by omitting coverage of some of the material. It can also be used as a two-semester course with the addition of some computer projects similar to those suggested herein. The book can also serve as a reference for engineers and scientists involved with pattern recognition, digital image processing, and artificial intelligence.

The author is indebted to Dale M. Grimes, Head of the Department of Electrical Engineering of The Pennsylvania State University, for his encouragement and support during the preparation of manuscript. The author also wishes to thank George J. McMurtry, Associate Dean of the College of Engineering at The Pennsylvania State University, for his valuable discussions and his generous permission to freely use some of his class notes in Chapters 3 and 4 and in the discussion of canonical analysis and its application to large data-set problems.

Sing-Tze Bow

Contents

Contents   *vii*

part I
PRINCIPLES OF PATTERN RECOGNITION

1
Introduction

1.1 PATTERNS AND PATTERN RECOGNITION

The patterns we encounter can fall into two categories: abstract and
concrete. Examples of abstract items include ideas and arguments.
Recognition of such patterns, termed *conceptual recognition*, belongs
to another branch of artificial intellegence and is beyond the scope of
this book.

Examples of concrete items include characters, symbols, pictures,
biomedical images, three-dimensional physical objects, target signatures, speech waveforms, electrocardiograms, electroencephalograms,
and seismic waves. Some of these items are spatial, whereas others
are temporal. In the last couple of decades, interest has focused on
two types of pattern recognition problems:

1. The mechanism of the pattern recognition system possessed
   by living organisms. Psychologists, physiologists, biologists,
   and neurophysiologists have devoted considerable effort
   toward exploring how living things perceive objects. Most of
   their results have been reported in the literature of bionics
   and other relevant disciplines.
2. The development of theories and techniques for computer
   implementation of a given recognition task. This is a subject
   that currently challenges both engineers and applied mathe-
   maticians. There is no unifying theory available that can be
   applied to all kinds of pattern recognition problems. Most
   techniques are problem oriented. Systematic presentation of
   the theories and techniques forms the basis of this book.

1.2 CONFIGURATION OF THE PATTERN RECOGNITION SYSTEM

1.2.1 Three Phases in Pattern Recognition

In pattern recognition we can divide an entire task into three phases: data acquisition, data preprocessing, and decision classification, as shown in Fig. 1.1. In the data acquisition phase, analog data from the physical world are gathered through a transducer and converted to digital format suitable for computer processing. In this stage, the physical variables are converted into a set of measured data, indicated in the figure by electric signals, $x(r)$, if the physical variables are sound (or light intensity) and the transducer is a microphone (or photocells). The measured data are then used as the input to the second phase (data preprocessing) and grouped into a set of characteristic features (x_N) as output. The third phase is actually a classifier which is in the form of a set of decision functions. With this set of features (x_N) the object may be classified. Figure 1.2 is a schematic diagram of an actual aerial multispectral scanner and data analysis system. The set of data at B, C, and D are in the pattern space, feature space, and classification space, respectively.

1.2.2 Representation of Pattern and Approaches to Their Machine Recognition

In Multidimensional Vector Form

As discussed in Sec. 1.2.1, there will be a set of collected, measured data after data acquisition. If the data to be analyzed are physical objects or images, the data acquisition device can be a television camera, a high-resolution camera, a multispectral scanner, or other device. For other types of problems, such as economic problems, the data acquisition system can be a data tape.

One function of data preprocessing is to convert a visual pattern into an electrical pattern or to convert a set of discrete data into a mathematical pattern so that those data are more suitable for computer analysis. The output will then be a pattern vector, which appears as a point in a pattern space.

To clarify this idea, let us take a simple visual image as the system input. If the image is scanned by a 12-channel multispectral scanner, we obtain, for a single picture point, 12 values, each corresponding to a separate spectral response. If the image is treated as a color image, three fundamental color-component values can be obtained, each corresponding, respectively, to a red, green, or blue spectrum band.

Each spectrum component value can be considered as a variable in n-dimensional space, known as *pattern space*, where each spectrum component is assigned to a dimension. Each pattern then appears as a point in the pattern space. It is a vector composed of n component

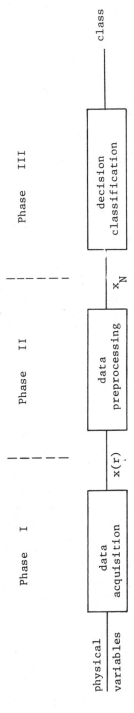

FIG. 1.1 Conceptual representation of a pattern recognition problem.

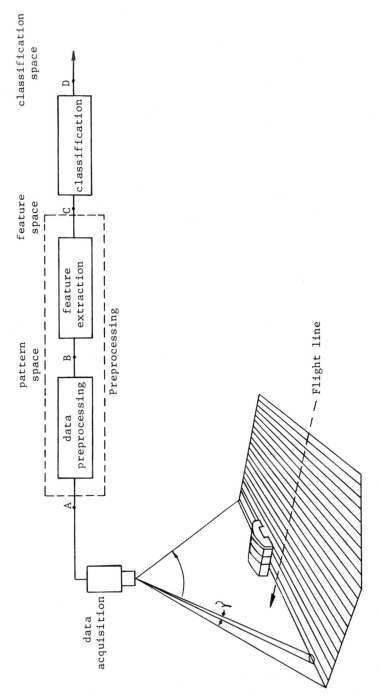

FIG. 1.2 Multispectral scanner and data analysis system.

values in the n-dimensional coordinates. A pattern $\underline{x}$ can then be represented as

$$\underline{x} = \begin{vmatrix} x_1 \\ x_2 \\ \cdot \\ \cdot \\ \cdot \\ x_n \end{vmatrix} \qquad (1.1)$$

where the subscript n represents the number of dimensions. If $n < 3$, the space can be illustrated graphically. Pattern space $\underline{X}$ may be described by a vector of m pattern vectors such that

$$\underline{X} = \begin{vmatrix} \underline{x}_1^T \\ \underline{x}_2^T \\ \cdot \\ \cdot \\ \cdot \\ \underline{x}_m^T \end{vmatrix} = \begin{vmatrix} x_{11} & x_{12} & \cdots & x_{1n} \\ x_{21} & x_{22} & \cdots & x_{2n} \\ \cdot & \cdot & & \cdot \\ \cdot & \cdot & & \cdot \\ \cdot & \cdot & & \cdot \\ x_{m1} & x_{m2} & \cdots & x_{mn} \end{vmatrix} \qquad (1.2)$$

where the superscript T after each vector denotes its transpose, and the $\underline{x}_i^T = (x_{i1}, x_{i2}, \ldots, x_{in})$, $i = 1, \ldots, m$, represent pattern vectors.

The objective of the feature extraction shown in Fig. 1.2 functions as the dimensionality reduction. It converts the original data to a suitable form (feature vectors) for use as input to the decision processor for classification. Obviously, the feature vectors represented by

$$\underline{x}_i^T = (x_{i1}, x_{i2}, \ldots, x_{ir}) \qquad i = 1, \ldots, m \qquad (1.3)$$

are in a smaller dimension (i.e., $r < n$).

The decision processor shown in Fig. 1.2 operates on the pattern vector and yields a classification decision. As we discussed before, pattern vectors are placed in the pattern space as "points," and patterns belonging to the same class will cluster together. Each cluster represents a distinct class, and clusters of points represent different classes of patterns. The decision classifier implemented with a set of decision function serves to define the class to which a particular pattern belongs.

The output of the decision processor will be in the classification space. It is M-dimensional if the input patterns are to be classified into M classes. For the simplest two-class problem, M equals 2; for aerial-photo interpretation, M can be 10 or more; and for alphabet recognition M equals 26. But for the case of Chinese character recognition, M can be more than 10,000. In such a case, other representations have to be used as supplements.

Both the preprocessor and the decision processor are usually selected by the user or designer. The coefficients (or weights) used in the decision processor are either calculated on the basis of complete a priori information of statistics of patterns to be classified, or are adjusted during a training phase. During the training phase, a set of patterns from a training set is presented to the decision processor, and the coefficients are adjusted according to whether the classification of each pattern is correct or not. This may then be called an *adaptive* or *training* decision processor. Note that most of the pattern recognition systems are not adaptive on-line; this is so only during the training phase. Also note that the preprocessing and decision algorithms should not be isolated from each other. Frequently, the preprocessing scheme has to be changed to make the decision processing more effective.

As discussed previously, a priori knowledge as to correct classification of some data vectors is needed in the training phase of the decision processor. Such data vectors are referred to as *prototypes* and are denoted as

$$
\underline{z}_k^m =
\begin{vmatrix}
z_{k1}^m \\
\cdot \\
\cdot \\
\cdot \\
z_{ki}^m \\
\cdot \\
\cdot \\
\cdot \\
z_{kn}^m
\end{vmatrix}
\quad
\begin{aligned}
& k = 1, 2, \ldots, M \\
& m = 1, 2, \ldots, N_k
\end{aligned}
$$

where $k = 1, 2, \ldots, M$ indexes the particular pattern class; $m = 1, 2, \ldots, N_k$ indicates the mth prototype of the class ω_k; and $i = 1, 2, \ldots, n$ indexes its component in the n-dimensional pattern vector. M, N_k, and n denote, respectively, the number of pattern classes; the number of prototypes in the kth class, ω_k; and the number of dimensions of the pattern vectors.

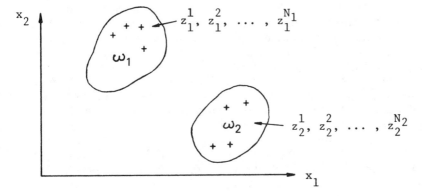

FIG. 1.3 Simple two-dimensional pattern space.

Prototypes from the same class share the same common properties and thus they cluster in a certain region of the pattern space. Figure 1.3 shows a simple two-dimensional pattern space. Prototypes $\underline{z}_1^1, \underline{z}_1^2, \ldots, \underline{z}_1^{N_1}$ cluster in ω_1; prototypes of another class, $\underline{z}_2^1, \underline{z}_2^2, \ldots, \underline{z}_2^{N_2}$, cluster in another region of the pattern space, ω_2. N_1 and N_2 are the number of prototypes in classes ω_1 and ω_2, respectively. The classification problem will simply be to find a separating surface that partitions the known prototypes into correct classes. This separating surface is expected to be able to classify the other unknown patterns if the same criterion is used in the classifier. Since patterns belonging to different classes will cluster into different regions in the pattern space, the distance metric between patterns can be used as a measure of similarity between patterns in the n-dimensional space.

Some conceivable properties between the distance metrics can be enumerated; thus

$$d(\underline{x},\underline{y}) = d(\underline{y},\underline{x})$$
$$d(\underline{x},\underline{y}) \leqslant d(\underline{y},\underline{z}) + d(\underline{z},\underline{x})$$
$$d(\underline{x},\underline{z}) \geqslant 0$$
$$d(\underline{x},\underline{y}) = 0 \quad \text{iff} \quad \underline{y} = \underline{x}$$

where $\underline{x}$, $\underline{y}$, and $\underline{z}$ are pattern vectors and $d(\cdot)$ denotes a distance function. Details regarding pattern classification by this approach are presented in subsequent chapters.

In Linguistically Descriptive Form

What we have just discussed is that each pattern is represented by a feature vector. The recognition of each pattern is usually made by

partitioning the feature space. This approach is commonly referred
to as the *decision theoretic approach.* The basis of this approach is
the meaningful representation of the data set in vector form. There
are, on the other hand, patterns whose structural properties are pre-
dominant in their descriptions. For such patterns, another approach,
called *syntactic pattern recognition,* will probably be more suitable.
The basis of this approach is to decompose a pattern into subpatterns
or primitives. The recognition of each pattern is usually made by
parisng the pattern structure according to syntax rules. Figure 1.4a
shows a simple pictorial pattern composed of a triangle and a pyramid.
Face F and triangle T are parts of object A. Triangles T_1 and T_2 are
parts of object B. The floor and wall together form the background of

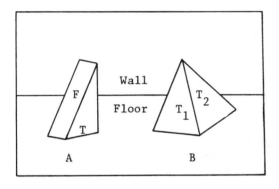

(a)

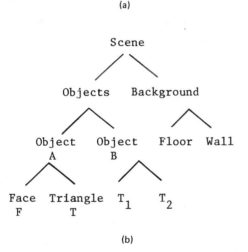

(b)

FIG. 1.4 Hierarchical representation of a simple scene.

the scene. Objects A and B together with the background constitute the whole scene, as shown in Fig. 1.4a. Figure 1.4b shows its hierarchical representation.

The image of the human chromosome is a good example of the use of syntactic description because of its strong structural regularity. There might be variations in the lengths of arms, but the basic shapes will be the same for certain types of chromosomes, such as submedian or telocentric ones. These variations can easily be recognized visually. Figure 1.5 shows the structural analysis of a submedian chromosome. Figure 1.5a shows bottom-up parsing on a submedian chromosome, Fig. 1.5b shows its structural representation, and Fig. 1.5c shows the primitives that we use for shape description.

When the boundary of the chromosome is traced in a clockwise direction, a submedian chromosome can be represented by a string such as abcbabdbabcbabdb if the symbols a, b, c, and d are suggested for the primitives shown in Fig. 1.5c. By the same token, a telocentric chromosome $\subseteq$ can be represented by ebabcbab. That is, a certain shape will be represented by a certain string of symbols. In the terminology of syntactic pattern recognition, a grammar, or set of rules of syntax, can be established for the generation of sentences for a certain type of chromosome. The sentences generated by two different grammars, say G_1 and G_2, will represent two different shapes; but the sentences generated by the same grammar, say G_1, represent the same category (e.g., submedian chromosomes), with tolerance for minor changes in shape proportion.

Chinese characters are another good example of the use of syntactic description. They were created and developed according to certain principles, such as functioning as pictophonemes and ideographs. They are composed of various primitives and possess strong structural regularities. With these regularities and semantics in mind, thousands of Chinese characters of complex configuration can be segregated and recombined. Thus the total amount of information will be greatly compressed. Thousands of complex ideographs can then be represented by a few semantic statements of morphological primitives. It can easily be seen that the total number of fundamental morphological primitives is far much less than 1000, and the complexities of the primitives are also much simpler than the original characters. It is possible, in the meantime, for "heuristics" to play an important role in pattern recognition and grammatical inference on these characters. In addition to structural description of the whole character, the structural approach has been applied to primitive description and extraction for Chinese character recognition.

1.2.3 Remarks on the Decision Theoretic Approach Versus the Syntactic Approach

Approaches for pattern recognition may be grouped into two categories: (1) the decision theoretic approach, and (2) the syntactic or structural

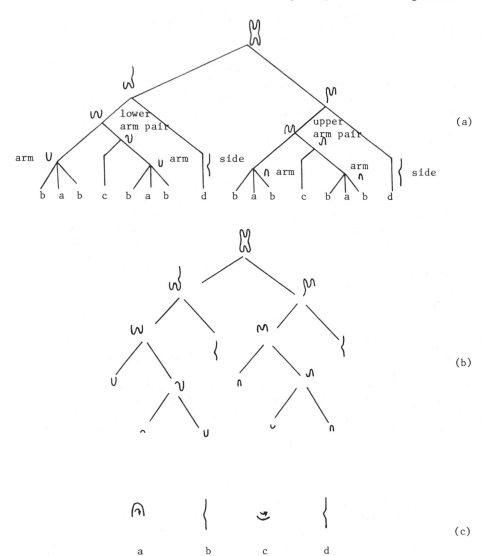

FIG. 1.5 Structural analysis of a submedian chromosome: (a) bottom-up parsing; (b) structural representation; (c) primitives used for the analysis.

approach. For some extreme problems the decision theoretic approach is most suitable, whereas for other extreme problems the syntactic approach is more suitable. The selection of approach depends primarily on the nature of the data set involved in a problem. For those problems where structural information is rich, it might be advantageous to use the syntactic method to show its power for problem description. If the data involved in the problem are better expressed in vector form and at the same time structural information about the patterns is not considered important, the decision theoretic method is recommended for its classification. There are many applications falling half way between these two extreme cases, and in such cases, the two approaches might complement each other. It might be easier or more helpful to use the decision theoretic method to extract some pattern primitives for the syntactic approach, particularly for noisy and distorted patterns. On the other hand, the syntactic method can help to give a structural picture instead of the mathematical results alone obtained by use of the decision theoretic approach. A comprehensive combination of these two approaches may result in an efficient and practical scheme for pattern recognition.

1.3 EXEMPLARY APPLICATIONS

The pattern recognition technique can be applied to more types of problems than can be enumerated. Readers should not feel restricted to the following applications, which are given for illustration only.

1.3.1 Weather Forecasting

In weather forecasting, the pressure contour map over a certain area (Fig. 1.6) constitutes the important data for study. From previous experience and a priori knowledge, several patterns (15 or more, depending on the area) can be specified on the sets of data maps. The

FIG. 1.6 Example of a pressure contour map over a certain area for weather forecasting studies.

weather forecasting problem then becomes to classify the existing
pressure contour patterns and to relate them to various weather conditions. Automatic and semiautomatic classifications by computer become
necessary when the number of maps builds up.

The two methods frequently used for pressure contour map classification are the correlation method and the principal component analysis (Karhunen-Loève) method. Both of these methods will give the
global features. Application of the syntactic method for weather forecasting problems, such as the use of string and/or tree representation
for pressure contour maps, is also under investigation.

1.3.2 Recognition of Handprinted Characters

Applications of handprinted character recognition are mainly for mail
sorting. This problem has been studied for a long time. Due to the
wide variations that exist in handwriting (see Fig. 1.7 for samples
printed by different persons), the correct recognition rate is still
not high enough for practical use.

Numerous approaches have been suggested for the recognition
of handprinted characters. So far, 121 constrained characters, including 52 uppercase and lowercase alphabetic letters, 10 numerals,
and other symbols are reported to be recognizable. Machine recognition of more sophisticated characters such as Chinese characters
is also under investigation.

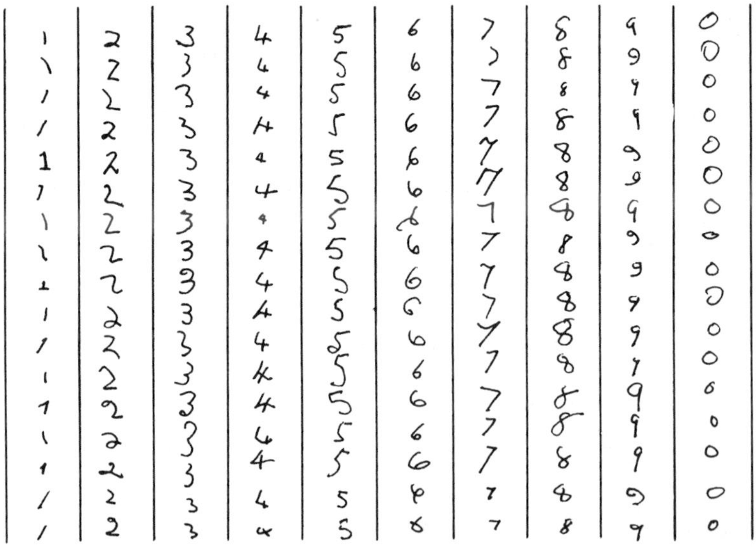

FIG. 1.7 Samples of handprinted numerals prepared by a variety of
people.

1.3.3 Speech Recognition

Speech recognition has numerous applications. One of these is its use
to supplement manual handling involved in mail sorting. When un-
sorted mail screened from the sorting line are more than manual control
operation can handle, speech recognition can be used as a supplementary
measure. The essentials of such methods are shown in Fig. 1.8.
Electrical signals converted from spoken words are first filtered
and sampled through tuned bandpass filters with center frequencies

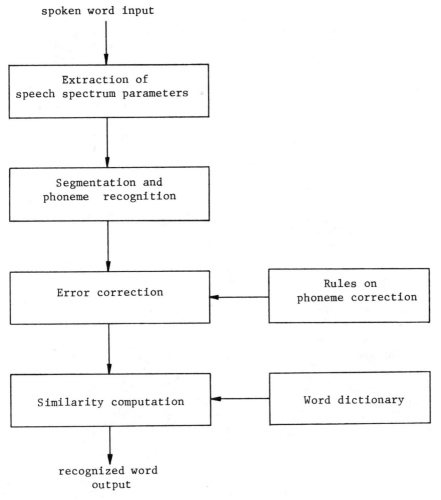

FIG. 1.8 Schematic diagram of a speech recognition system.

from 200 to 7500 Hz. Several specific parameters, such as spectral
local peaks, speech power, and those representing the gross pattern
of the spectrum, are extracted for segmentation and phoneme recognition. Errors that have occurred during segmentation and phoneme
recognition are corrected by means of preset phoneme correction rules,
and then similarity computation is carried out and words of maximum
similarity chosen for the solution.

1.3.4 Medical Analysis

Occupational diseases cause workers considerable concern as to job
selection. Early cures for such diseases depend on early and accurate
diagnosis. An example is coal miners' pneumoconiosis, a disease of
the lungs caused by continual inhalation of irritant mineral or metallic
particles. The principal symptom is the descent of the pulmonary
arteries (see Fig. 1.9 for an abnormal chest x-ray of a patient). Accurate diagnosis depends on accurate discrimination of the small
opacities of different types from the normal pulmonary vascularity pattern. These opacities appear here and there, sometimes in the interrib

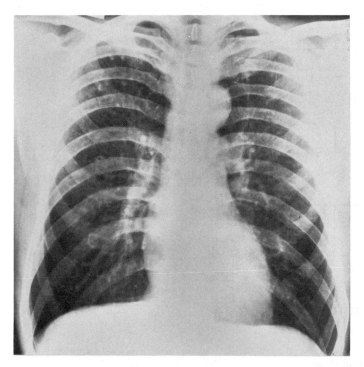

FIG. 1.9 Chest x-ray of a pneumoconiosis patient. (Courtesy of C. C.
Li, Department of Electrical Engineering, University of Pittsburgh.)

space and sometimes in the rib spaces. Those appearing in the rib spaces, overlapped by shadows cast by the major pulmonary arteries, are very hard to recognize. Pattern recognition technique can usefully be applied to this kind of problem.

To perform this task, the chest x-ray has to be processed to eliminate the major pulmonary arteries, the rib contours, and so on, to provide a frame of reference for the suspicious objects detected. The differences in various texture features are used to classify coal miners' chest x-rays into normal and abnormal classes. Four major categories have been established to indicate the severity of the disease according to the profusion of opacities in the lung region.

1.3.5 Satellite and Aerial-Photo Interpretation

Satellite and/or aerial images are used for both military and civil purposes. Among the civil applications, the remote sensing of earth resources either on or under the surface of the earth is an important topic for study. Remote sensing has a wide variety of applications: in agriculture, forestry, city planning, geology, geography, and railway line exploitation. The data received from the satellite or from the tape recorded during airplane flight is first restored and enhanced in image form, and then interpreted by a specialist. The principal disadvantage with visual interpretation lies in the extensive training and intensive labor required. In addition, visual interpretation cannot always fully evaluate spectral characteristics. This is because of the limited ability of the eye to discern tonal values on an image and the difficulty an interpreter has in analyzing numerous spectral images simultaneously. In applications where spectral patterns are highly informative, it is therefore preferable to analyze numerical rather than pictorial image data. For these reasons, computer data processing and pattern classification will play an increasingly important role in such applications. Both temporal and spatial patterns are studied to meet different problem requirements. Details of these applications will not be presented here, as a more detailed worked-out problem is given later to illustrate some of the principles discussed in Chap. 4.

2
Nonparametric Decision Theoretic Classification

Pattern recognition algorithms are often classified as either parametric or nonparametric. For some classification tasks, pattern categories are known a priori to be characterized by a set of parameters. A parametric approach is to define the discriminant function by a class of probability densities defined by a relatively small number of parameters.

There exist many other classifications in which no assumptions can be made about the characterizing parameters. Nonparametric approaches are designed for those tasks. Although some parametrized discriminant functions (e.g., the coefficients of a multivariate polynomial of some degree) are used in nonparametric methods, no conventional form of the distribution is assumed. This differs from the parametric approach. In the parametric approach, pattern classes are usually assumed to arise from a multivariate gaussian distribution where the parameters are the mean and covariances.

In this chapter the emphasis is on a discussion of non parametric decision theoretic classification. Based on the nature of the problem, we are going to discuss several related topics in succession. To start, some technical definitions of decision surfaces and discriminant functions are introduced, and then the discussion is directed to the general form of the discriminant function, its properties and classifier, based on this sort of discriminant function. Cases dealing with linear discriminant functions, piecewise linear discriminant functions, and nonlinear discriminant functions are discussed separately. A discussion of ϕ machines and their capacity to classify patterns follows to generalize the nonparametric decision theoretic classification method. At the end of the chapter, potential functions used as discriminant functions are included to give a more complete description of the nonparametric decision theoretic classification.

2.1 DECISION SURFACES AND DISCRIMINANT FUNCTIONS

As mentioned in Chap. 1, each pattern appears as a point in the pattern space. Patterns pertaining to different classes will fall into different regions in the pattern space (i.e., different classes of patterns will cluster in different regions and can easily be separated by separating surfaces). Separating surfaces, called *decision surfaces,* can formally be defined as surfaces in n dimensions which are used to separate known patterns into their respective categories and are used to classify unknown patterns. Such decision surfaces are called *hyperplanes* and are $(n - 1)$-dimensional. When $n = 1$, the decision surface is a point, as shown in Fig. 2.1, where point A is the point separating classes ω_1 and ω_2, and point B is the separating point between ω_2 and ω_3. When $n = 2$, the decision surface becomes a line,

$$w_1 x_1 + w_2 x_2 + w_3 = 0 \tag{2.1}$$

as shown in Fig. 2.2 When $n = 3$, the surface is a plane. When $n = 4$ or higher, the decision surface is a hyperplane represented by

$$w_1 x_1 + w_2 x_2 + w_3 x_3 + \cdots + w_n x_n + w_{n+1} = 0 \tag{2.2}$$

expressed in matrix form as

$$\underline{w} \cdot \underline{x} = 0 \tag{2.3}$$

where

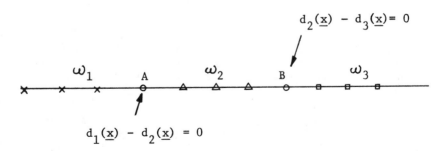

FIG. 2.1 One-dimensional pattern space.

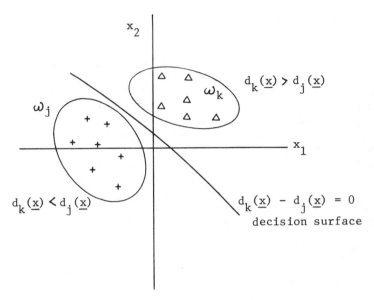

FIG. 2.2 Two-dimensional pattern space.

$$\underline{w} = \begin{bmatrix} w_1 \\ w_2 \\ \cdot \\ \cdot \\ \cdot \\ w_n \\ w_{n+1} \end{bmatrix} \quad \text{and} \quad \underline{x} = \begin{bmatrix} x_1 \\ x_2 \\ \cdot \\ \cdot \\ \cdot \\ x_n \\ 1 \end{bmatrix}$$

$\underline{w}$ and $\underline{x}$ are called the augmented weight vector and the augmented pattern vector, respectively. The scalar term w_{n+1} has been added to the weight function for coordinate translation purposes. To make the equation a valid vector multiplication, the input vector $\underline{x}$ has been augmented to become $(n + 1)$-dimensional by adding $x_{n+1} = 1$. This will allow a translation of all linear discriminant functions to pass through the origin of the augmented space when desired.

A discriminant function is a function $d(\underline{x})$ which defines the decision surface. As shown in Fig. 2.2, $d_k(\underline{x})$ and $d_j(\underline{x})$ are values of the discriminant function for patterns $\underline{x}$ in classes k and j, respectively. $d(\underline{x}) = d_k(\underline{x}) - d_j(\underline{x}) = 0$ will be the equation defining the surface that separates classes k and j. We can then say that

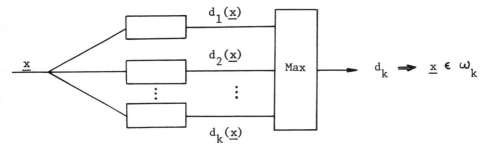

FIG. 2.3 Schematic diagram of a simple classification system.

$$d_k(\underline{x}) > d_j(\underline{x}) \quad \forall \quad \underline{x} \in \omega_k$$

$$\forall \; j \neq k, \; j = 1, 2, \ldots, M \tag{2.4}$$

A system can then be built to classify pattern $\underline{x}$ as shown in Fig. 2.3. For a two-class problem,

$$d_1(\underline{x}) = d_2(\underline{x}) \tag{2.5}$$

or

$$d(\underline{x}) = d_1(\underline{x}) - d_2(\underline{x}) = 0 \tag{2.6}$$

will define the separating hyperplane between the two classes.

In general, if we have M different classes, there will be $M(M - 1)/2$ separating surfaces. But some of the separating surfaces are redundant: only $M - 1$ are needed to separate the M classes. Figure 2.4 shows the number of separating surfaces as a function of the number of categories to be classified in a two-dimensional pattern space. From Fig. 2.4a we can see that the decision surface separating the two categories is a line. On the line, $d(\underline{x}) = 0$; below the separating line, $d(\underline{x}) > 0$; and above the line, $d(\underline{x}) < 0$. Thus the line $d(\underline{x}) = 0$ separates two different classes. Similarly, Fig. 2.4b is self-explanatory. Note that in the crosshatched region where $d_1(\underline{x}) < 0$ and $d_2(\underline{x}) < 0$, patterns belong neither to ω_1 nor to ω_2, and may be classified as ω_3. The same thing happens in Fig. 2.4c: patterns falling in the crosshatched portion of the plane do not belong to category 1, 2, or 3, and thus form a new category. Portions not mentioned in this pattern space are indeterminate regions.

Example For a two-class problem $(M = 2)$, find a discriminant function to classify the following two patterns into two categories:

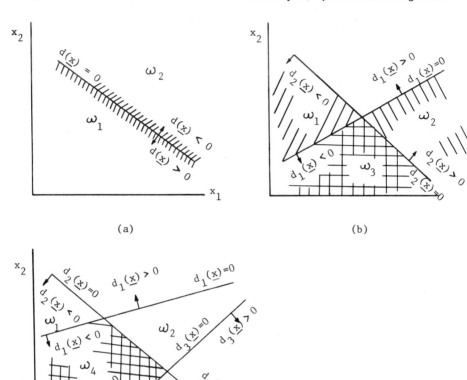

FIG. 2.4 Separating surfaces in a two-dimensional pattern space.

(a)

(b)

(c)

$$\underline{x}_1 = \begin{vmatrix} 1 \\ 4 \end{vmatrix} \quad \text{and} \quad \underline{x}_2 = \begin{vmatrix} 4 \\ 2 \end{vmatrix}$$

Try $d(\underline{x}) = x_1 - (1/2)x_2 - 2$ and see whether it can be used as the separating line for these two patterns. Substituting the augmented vectors of $\underline{x}_1$ and $\underline{x}_2$ into $d(\underline{x})$, we find that

$$d(\underline{x}_1) = \underline{w} \cdot \underline{x}_1 = \left(1 - \frac{1}{2} - 2\right)\begin{pmatrix} 1 \\ 4 \\ 1 \end{pmatrix} = -3 < 0$$

$$d(\underline{x}_2) = \underline{w} \cdot \underline{x}_2 = \left(1 - \frac{1}{2} - 2\right)\begin{pmatrix} 4 \\ 2 \\ 1 \end{pmatrix} = +1 > 0$$

Then the pattern $\underline{x}_1$ can be classified in one category and $\underline{x}_2$ in another category according to this discriminant function.

2.2 LINEAR DISCRIMINANT FUNCTIONS

As mentioned in Sec. 2.1, patterns falling in different regions in the pattern space can be grouped into different categories by means of separating surfaces which are defined by discriminant functions. The discriminant function may be linear or nonlinear according to the nature of the problem.

In this section we start by discussing the general form of the linear discriminant function and then apply it to the design of the minimum distance classifier. Linear separability will also be discussed.

2.2.1 General Form

The linear discriminant function will be of the following form:

$$d_k(\underline{x}) = w_{k1}x_1 + w_{k2}x_2 + \cdots + w_{kn}x_n + w_{k,n+1}x_{n+1} \qquad (2.7)$$

Put in matrix form,

$$d_k(\underline{x}) = \underline{w}_k^T \underline{x} \qquad (2.8)$$

where

$$\underline{w}_k = \begin{bmatrix} w_{k1} \\ w_{k2} \\ \cdot \\ \cdot \\ \cdot \\ w_{kn} \\ w_{k,n+1} \end{bmatrix} \qquad \underline{x} = \begin{bmatrix} x_1 \\ x_2 \\ \cdot \\ \cdot \\ \cdot \\ x_n \\ 1 \end{bmatrix}$$

and $x_{n+1} = 1$ in the augmented $\underline{x}$ pattern vector. For a two-class problem, $M = 2$,

$$d(\underline{x}) = \underline{w}_1^T \underline{x}_1 - \underline{w}_2^T \underline{x}_2 = (\underline{w}_1 - \underline{w}_2)^T \underline{x} = 0 \tag{2.9}$$

which is a hyperplane passing through the origin in an augmented feature space, for the reason discussed previously.

2.2.2 Minimum Distance Classifier

Although it is one of the earliest methods suggested, the minimum distance classifier is still an effective tool in solving the pattern classification problem. The decision rule used in this method is

$$\underline{x} \in \omega_j \quad \text{if } D(\underline{x}, \underline{z}_j) = \min_k D(\underline{x}, \underline{z}_k) \quad k = 1, \ldots, M \tag{2.10}$$

where $D(\cdot)$ is a metric called the *euclidean distance* of an unknown pattern $\underline{x}$ from $\underline{z}_k$, and $\underline{z}_k$ is the prototype average (or class center) for class ω_k. Then

$$D(\underline{x}, \underline{z}_k) = |\underline{x} - \underline{z}_k| \tag{2.11}$$

Remembering that if

$$D(\underline{x}, \underline{z}_k) > D(\underline{x}, \underline{z}_j) \quad \forall \quad j, k \tag{2.12}$$

then

$$D^2(\underline{x}, \underline{z}_k) > D^2(\underline{x}, \underline{z}_j) \tag{2.13}$$

is true for most cases. In other words, we can use D^2 to replace D in the decision rule above. Then we have

$$D^2(\underline{x}, \underline{z}_k) = |\underline{x} - \underline{z}_k|^2 \tag{2.14}$$

Put in matrix form,

$$D^2(\underline{x}, \underline{z}_k) = (\underline{x} - \underline{z}_k)^T (\underline{x} - \underline{z}_k)$$
$$= \underline{x}^T \underline{x} - 2\underline{x}^T \underline{z}_k + \underline{z}_k^T \underline{z}_k \tag{2.15}$$

after expanding. On the right-hand side of the expression above, $\underline{x}^T\underline{x}$ is constant for all k, and therefore can be eliminated. Thus, to seek the minimum of $D(\underline{x}, \underline{z}_k)$ is equivalent to seeking

$$\min_{k} [-2 \underline{x}^T \underline{z}_k + \underline{z}_k^T \underline{z}_k] \qquad (2.16)$$

or alternatively, to seeking

$$\max_{k} [\underline{x}^T \underline{z}_k - \frac{1}{2} \underline{z}_k^T \underline{z}_k] \qquad k = 1, 2, \ldots, M \qquad (2.17)$$

which is the decision rule for the minimum distance classifier. The discriminant function used in the classifier can then be expressed as

$$d_k(\underline{x}) = \underline{x}^T \underline{z}_k - \frac{1}{2} \underline{z}_k^T \underline{z}_k = \underline{x}^T \underline{z}_k - \frac{1}{2} |\underline{z}_k|^2 = \underline{x}^T \underline{w} \qquad (2.18)$$

where

$$\underline{w} = \begin{bmatrix} z_k^1 \\ z_k^2 \\ z_k^3 \\ \cdot \\ \cdot \\ \cdot \\ z_k^n \\ -\frac{1}{2} |z_k|^2 \end{bmatrix}$$

and $\underline{x}$ is an augmented pattern vector. Note that the decision surface between any two classes ω_i and ω_j is formed by the perpendicular bisectors of $\underline{z}_i - \underline{z}_j$ shown by dashed lines in Fig. 2.5. The proof for this is not difficult. From Eq. (2.18), the decision surface between $\underline{z}_1$ and $\underline{z}_2$ is

$$d(\underline{x}) = \underline{x}^T (\underline{z}_1 - \underline{z}_2) - \frac{1}{2} (\underline{z}_1^T \underline{z}_1 - \underline{z}_2^T \underline{z}_2) = 0 \qquad (2.19)$$

Obviously, the midpoint between $\underline{z}_1$ and $\underline{z}_2$ is on the decision surface. This can be shown by direct substitution of this midpoint into Eq. (2.19), which shows that the equation is satisfied. Similarly, we can find that all other points on the boundary surface (a line in this case) also satisfy Eq. (2.19). It can also be proved that vector $(\underline{z}_1 - \underline{z}_2)$ is in the same direction as the unit normal to the hyperplane, which is

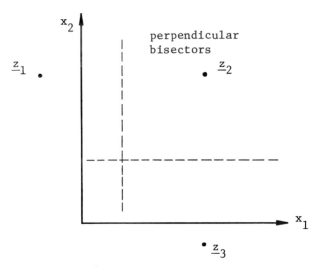

FIG. 2.5 Geometrical properties of the decision surfaces.

$$\frac{z_1 - z_2}{|z_1 - z_2|} \tag{2.20}$$

Note also that the minimum distance classifier (MDC) uses a single point to represent each class. This representation would be all right if the class were normally distributed with equal variances in all directions, as shown in Fig. 2.6a. But if the class is not normally distributed with equal variances in all directions, as shown in Fig. 2.6b, misclassification will occur. Even if $D_1 < D_2$, point + should be classified to ω_2 instead of ω_1. Similarly, in Fig. 2.6c, point + might be classified in class ω_3 by the MDC, but it is really closer to ω_2. That is, single points do not represent classes ω_1, ω_2, and ω_3 very well.

If each pattern category is represented by multiprototypes instead of a single prototype, then

$$D(\underline{x}, \omega_k) = \min_{m=1,\ldots,N_k} [D(\underline{x}, \underline{z}_k^m)] \tag{2.21}$$

where k represents the kth category, m represents the mth prototype, and N_k represents the number of prototypes used to represent category k. Equation (2.21) gives the smallest of the distances between $\underline{x}$ and each of the prototypes of ω_k. The decision rule then becomes

$$\underline{x} \in \omega_j \quad \text{if } D(\underline{x}, \omega_j) = \min_{k=1,\ldots,M} D(\underline{x}, \omega_k) \tag{2.22}$$

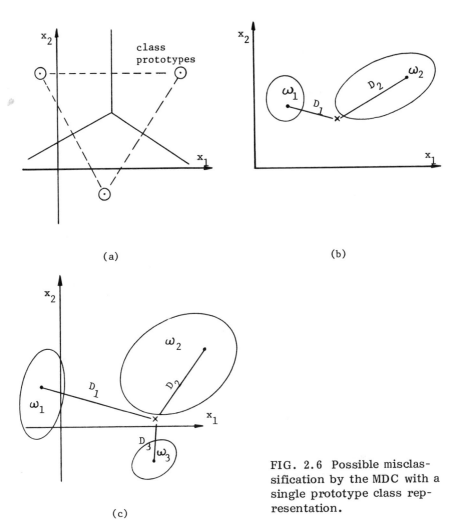

(a)   (b)

(c)

FIG. 2.6 Possible misclassification by the MDC with a
single prototype class representation.

where $D(\underline{x}, \omega_j)$ is given by Eq. (2.21). The discriminant function
changes correspondingly to the following form:

$$d_k(\underline{x}) = \max_{m=1,\ldots,N_k} [\underline{x}^T \underline{z}_k^m - \frac{1}{2} |\underline{z}_k^m|^2] \quad k = 1, 2, \ldots, M \quad (2.23)$$

2.2.3 Linear Separability

Some properties relating to the classification, such as linear separability of patterns, are discussed next. Pattern classes are said to

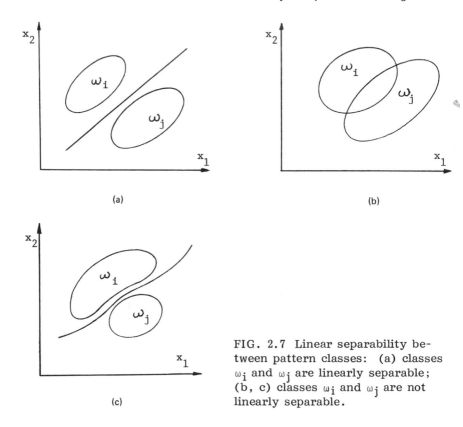

(a)

(b)

(c)

FIG. 2.7 Linear separability between pattern classes: (a) classes ω_i and ω_j are linearly separable; (b, c) classes ω_i and ω_j are not linearly separable.

be *linearly separable* if they are classifiable by any linear function, as shown in Fig. 2.7a, whereas the classes in Fig. 2.7b and c are not classifiable by any function.

From Fig. 2.7 we can see that the decision surfaces in linearly separable problems are convex. By definition, a function is said to be

(a)

(b)

FIG. 2.8 Convexity property of a function: (a) convex decision surface; (b) not convex.

convex in a given region if a straight line drawn within that region lies entirely in or above the function. The regional function shown in Fig. 2.8a is said to be convex, since straight lines ab and ac are all above the function curve, whereas that shown in Fig. 2.8b is not.

2.3 PIECEWISE LINEAR DISCRIMINANT FUNCTIONS

So far, the discussion has focused on linear separable problems. It does not seem to be difficult to deal with such problems. But for linearly nonseparable problems, more effective approaches must be sought. This will be the topic of the next several sections.

2.3.1 Definition and Nearest Neighbor Rule

A piecewise linear function is a function that is linear over subregions of the feature space. These piecewise linear discriminant functions give piecewise linear boundaries between categories as shown in Fig. 2.9a, where the boundary surface shown in Fig. 2.9b is nonconvex. The discriminant functions are given by

$$d_k(\underline{x}) = \max_{m=1,\ldots,N_k} [d_k^m(\underline{x})] \quad k = 1, \ldots, M, \qquad (2.24)$$

that is, we find the maximum $d_k^m(\underline{x})$ among the prototypes of class k, where N_k is the number of prototypes in class k and

$$d_k^m(\underline{x}) = w_{k1}^m x_1 + w_{k2}^m x_2 + \cdots + w_{kn}^m x_n + w_{k,n+1}^m$$

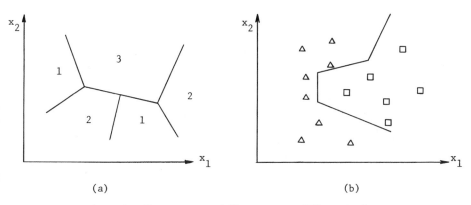

(a)　　　　　　　　　　　　(b)

FIG. 2.9 Piecewise linear separability among different classes.

$$= (\underline{w}_k^m)^T \underline{x} \qquad\qquad (2.25)$$

where

$$\underline{w}_k^m \begin{bmatrix} w_{k1}^m \\ w_{k2}^m \\ \cdot \\ \cdot \\ \cdot \\ w_{kn}^m \\ w_{k,n+1}^m \end{bmatrix}$$

and

$$\underline{x} = \begin{bmatrix} x_1 \\ x_2 \\ \cdot \\ \cdot \\ \cdot \\ x_n \\ 1 \end{bmatrix}$$

Three different cases of the pattern classification problem can be enumerated:

Case 1: Each pattern class is separable from all other classes by a single decision surface, as shown in Fig. 2.10a, where several indeterminate regions can be seen to exist.

Case 2: Each pattern class is pairwise separable from every other class by a distinct decision surface. Indeterminate regions may also exist. In this case, no class is separable from the others by a single decision surface. For example, ω_1 can be separated from ω_2 by the surface $d_{12}(\underline{x}) = 0$ and from ω_3 by $d_{13}(\underline{x}) = 0$ (see Fig. 2.10b). There will be $M(M - 1)/2$ decision surfaces which are represented by

$$d_{ij}(\underline{x}) = 0 \qquad\qquad (2.26)$$

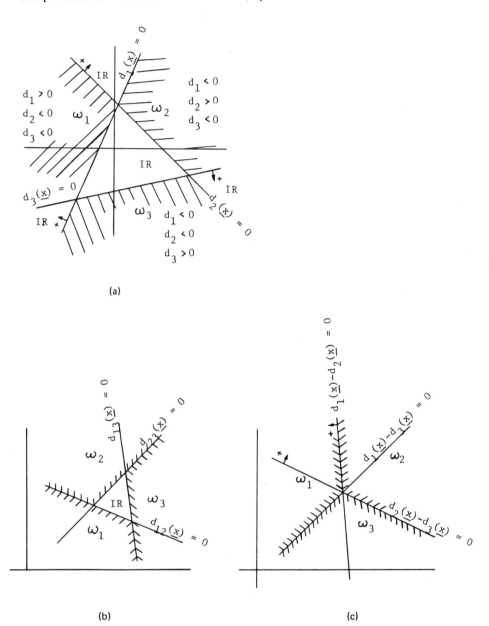

FIG. 2.10 Three different cases in pattern classification: (a) each class is separable from others by a single decision surface; (b) each class is pairwise separable from others by a distinct decision surface; (c) same as part (b) but with no indeterminate regions.

and

$$\underline{x} \in \omega_i \quad \text{when} \quad d_{ij} > 0 \quad \forall \quad j \neq i \tag{2.27}$$

Case 3: Each pattern class is pairwise separable from every other class by a distinct decision surface, but with no indeterminate regions (a special case of case 2); see Fig. 2.10c. In this case, there are M decision functions and

$$d_k(\underline{x}) = (\underline{w}_k)^T \underline{x} \quad k = 1, 2, \ldots, M \tag{2.28}$$

and

$$\underline{x} \in \omega_i \quad \text{if} \quad d_i(\underline{x}) > d_j(\underline{x}) \quad \forall \quad j \neq i \tag{2.29}$$

2.3.2 Layered Machines

A two-layered machine is also known as a *committee machine,* so named because it takes a fair vote for each linear discriminant function output to determine the classification. That part to the left of the Σ unit shown on Fig. 2.11 is the first layer and that to the right of it is the second layer. $\underline{w}_1, \underline{w}_2, \ldots, \underline{w}_R$ are the different n-dimensional weight vectors for each discriminant functions and are, respectively,

$$\underline{w}_1^T = (w_{11}, w_{12}, \ldots, w_{1n})$$

$$\underline{w}_2^T = (w_{21}, w_{22}, \ldots, w_{2n}) \tag{2.30}$$

$$\cdot$$
$$\cdot$$
$$\cdot$$

$$\underline{w}_R^T = (w_{R1}, w_{R2}, \ldots, w_{Rn})$$

The first layer consists of an odd number of linear discriminant surfaces whose outputs are clipped by the threshold logic unit as +1 or −1, depending on the value of $f(\underline{x})$ to describe on which half of the feature space the particular pattern input falls. The second layer is a single linear surface with unity weight vector used to decide to which class the particular pattern will finally be assigned.

When an adaptive loop is placed in the threshold logic unit for the training of $\underline{w}$, the threshold logic unit is called an *adaptive linear threshold element* (ADALINE). When multiple ADALINEs are used in the machine, it is called a MADALINE—a comittee machine.

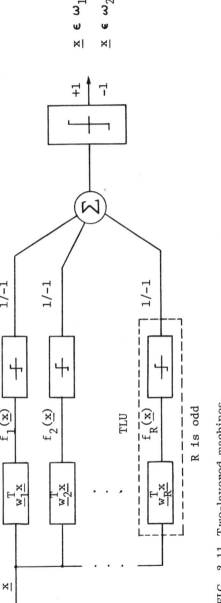

FIG. 2.11 Two-layered machines.

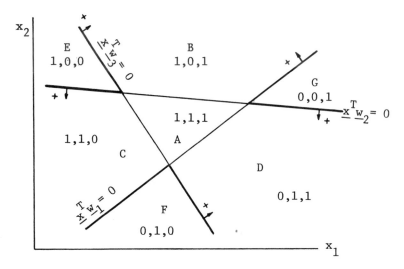

FIG. 2.12 Geometrical interpretation of a simple two-class classification problem by a layered machine.

Let us take a simple two-class problem to interpret the machine geometrically. Suppose that we have three threshold logic units in the first layer; $R = 3$. $\underline{w}_1^T \underline{x} = 0$, $\underline{w}_2^T \underline{x} = 0$, and $\underline{w}_3^T \underline{x} = 0$ will define three hyperplanes in the feature space, shown in Fig. 2.12.

The layered machine will divide the pattern space geometrically into seven regions. Table 2.1 lists values of $\underline{w}_1^T \underline{x}$, $\underline{w}_2^T \underline{x}$, and $\underline{w}_3^T \underline{x}$ in

TABLE 2.1 Values of $\underline{w}_1^T \underline{x}$, $\underline{w}_2^T \underline{x}$, and $\underline{w}_3^T \underline{x}$ in Different Regions of the Pattern Space

$\underline{w}^T \underline{x}$	Subregion							
	A	B	C	D	E	F	G	
$\underline{w}_1^T \underline{x}$	1	1	1	0	1	0	0	0
$\underline{w}_2^T \underline{x}$	1	0	1	1	0	1	0	0
$\underline{w}_3^T \underline{x}$	1	1	0	1	0	0	1	0
$\underline{x} \ \epsilon$	ω_1	ω_1	ω_1	ω_1	ω_2	ω_2	ω_2	—

each region, with 1's and 0's denoting greater than and less than zero, respectively. No regions lie on the negative side of all three hyperplanes, so 0, 0, 0 can never occur.

2.4 NONLINEAR DISCRIMINANT FUNCTIONS

The linear discriminant function is the simplest discriminant function, but often the nonlinear discriminant function should be used. Quadratic discriminant functions have the following form:

$$d(\underline{x}) = \sum_{j=1}^{n} w_{jj}x_j^2 + \sum_{j=1}^{n-1} \sum_{k=j+1}^{n} w_{jk}x_j x_k + \sum_{j=1}^{n} w_j x_j + w_{n+1} \qquad (2.31)$$

The first set of weights on the right-hand side of Eq. (2.31), w_{jj}, $j = 1, 2, \ldots, n$, consists of n weights; the second set, w_{jk}, $j = 1, 2, \ldots, n - 1$, $k = 1, 2, \ldots, n$, consists of $n(n - 1)/2$ weights; the third set, w_j, $j = 1, 2, \ldots, n$, n weights; and the last set w_{n+1}, only one weight. Hence the total number of weights on $d(\underline{x})$ is $(n + 1)$ $(n + 2)/2$. When expression (2.31) is put in matrix form,

$$d(\underline{x}) = \underline{x}^T A \underline{x} + \underline{x}^T B + C \qquad (2.32)$$

where

$$A = \begin{vmatrix} w_{11} & w_{12} & \cdots & w_{1n} \\ w_{21} & w_{22} & & w_{2n} \\ \cdot & & & \cdot \\ \cdot & & & \cdot \\ \cdot & & & \cdot \\ w_{n1} & w_{n2} & \cdots & w_{nn} \end{vmatrix}$$

$$B = \begin{vmatrix} w_1 \\ w_2 \\ \cdot \\ \cdot \\ \cdot \\ w_n \end{vmatrix} \qquad C = w_{n+1}$$

Note that if all the eigenvalues λ of A are positive, the quadratic form $\underline{x}^T A \underline{x}$ will never be negative, and

$$\underline{x}^T A \underline{x} = 0 \quad \text{iff} \quad \underline{x} = 0 \tag{2.33}$$

That means that matrix A is positive definite and the quadratic form is also positive definite. But if one or more eigenvalues (λ's) equal zero while the others are positive, matrix A and the quadratic $\underline{x}^T A \underline{x}$ are positive semidefinite.

Remember that on the decision surface, $d_i(\underline{x}) = d_j(\underline{x})$. In other words, the decision surface is defined by $d_i(\underline{x}) - d_j(\underline{x}) = 0$. For the quadratic case, the quadratic decision surface is given by an equation of the form

$$\underline{x}^T [A_i - A_j] \underline{x} + \underline{x}^T [B_i - B_j] + [C_i - C_j] = 0 \tag{2.34}$$

Varieties of the quadratic surfaces can be defined for different values of A, which is equal to $A_i - A_j$.

If A is positive definite, the decision surface is a hyperellipsoid with axes in the direction of the eigenvectors of A. If A = aI, an identity matrix, the decision surface is a hypersphere. If A is positive semidefinite, the decision surface is a hyperellipsoidal cylinder whose cross sections are lower-dimensional hyperellipsoidals with axes in the direction of eigenvectors of A of nonzero eigenvectors. Otherwise (i.e., when none of the conditions above is fulfilled by A or A is negative definite), the decision surface is a hyperhyperboloid.

2.5 ϕ MACHINES

2.5.1 Formulation

ϕ machines are a kind of classification system in which ϕ functions are used for pattern classification. The ϕ function (or generalized decision function) is a discriminant function that can be written in the form

$$d(\underline{x}) = \phi(\underline{x}) = w_1 f_1(\underline{x}) + w_2 f_2(\underline{x}) + \cdots + w_M f_M(\underline{x}) + w_{M+1} \tag{2.35}$$

where the $f_i(\underline{x})$, $i = 1, \ldots, M$, are linearly independent, real, and single-valued functions which are independent of the w_i (weights).

Note that $\phi(\underline{x})$ is linear with respect to w_i, but the $f_i(\underline{x})$ are not necessarily assumed to be linear. There are M + 1 degrees of freedom in this system. The same nonlinear discriminant function problem as used in Sec. 2.4 is taken for illustration.

$$d(x) = \sum_{j=1}^{n} w_{jj} x_j^2 + \sum_{j=1}^{n-1} \sum_{k=j+1}^{n} w_{jk} x_j x_k + \sum_{j=1}^{n} w_j x_j + w_{n+1}$$

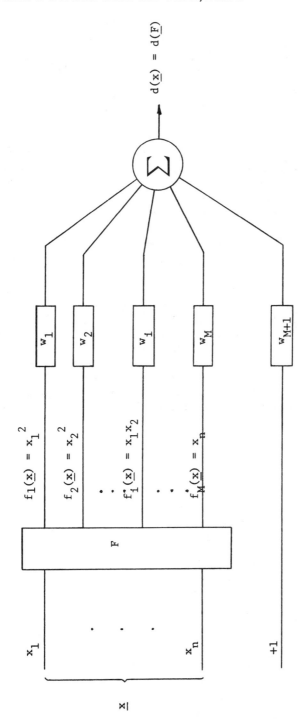

FIG. 2.13 ϕ machines.

$$= w_{11}x_1^2 + w_{22}x_2^2 + \cdots + w_{12}x_1x_2 + w_{13}x_1x_3$$

$$+ \cdots + w_{n-1,n}x_{n-1}x_n$$

$$+ w_1x_1 + w_2x_2 + \cdots + w_nx_n + w_{n+1} \tag{2.36}$$

A schematic diagram of the ϕ machine for this problem is shown in Fig. 2.13. The F block is a quadratic processor and $\underline{F} = (f_1, f_2, f_3, \ldots, f_M)$. The first n components of $\underline{F}$ are $x_1^2, x_2^2, \ldots, x_n^2$; the next $n(n-1)/2$ components are all the pairs $x_1x_2, x_1x_3, \ldots, x_{n-1}x_n$; and the last n components are $x_1, x_2, \ldots, x_n$. The total number of the components is $M = n(n+3)/2$. We have then transformed from an n-dimensional pattern space to an M-dimensional ϕ space.

2.5.2 Capacity of ϕ Machines for Classifying Patterns

Let us compute the number of dichotomies that can be obtained from N patterns. Assume that $M = 2$ and there are N n-dimensional patterns. Since each pattern may fall either in ω_1 or ω_2, there are 2^N distinct ways in which these N patterns could be dichotomized. For $N = 3$ we will have eight dichotomies, and for $N = 4$ we have 16 dichotomies. The total number of dichotomies that a linear discriminant function (ϕ machine in ϕ space) can affect is dependent only on n and N, not on how the patterns lie in the pattern space in the form of the ϕ function.

Let $D(N,n)$ be the number of dichotomies that can be affected by a linear machine (linear dichotomies) on N patterns in n-dimensional space. In the four-pattern example given in Fig. 2.14, ℓ_2 dichotomizes

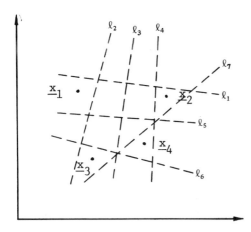

FIG. 2.14 Linear dichotomization of four patterns.

x_1 from x_2, x_3, and x_4; ℓ_5 dichotomizes x_1 and x_2 from x_3 and x_4; ℓ_7 dichotomizes x_4 from x_1, x_2, and x_3; ℓ_3 dichotomizes x_1 and x_3 from x_2 and x_4; and ℓ_4 dichotomizes x_2 from x_1, x_3, and x_4. That is, in the problem we have here —N = 4 and n = 2—we have seven linear dichotomies. Each of them can divide the patterns in either one of two ways, such as in the dichotomy by ℓ_3:

$$x_1, x_3 \in \omega_1 \quad x_2, x_4 \in \omega_2$$

or

$$x_1, x_3 \in \omega_2 \quad x_2, x_4 \in \omega_1$$

The number of linear dichotomies of N points in n-dimensional euclidean space is equal to twice the number of ways in which the points can be partitioned by an (n − 1)-dimensional hyperplane; so

$$D(4,2) = 2 \cdot 7 = 14$$

Comparing the total number of dichotomies, $2^N = 16$, we find that two of these are not linearly implementable. It is not difficult to see that x_1 and x_4 cannot be linearly separated from x_2 and x_3.

In general, for a set of N points in an n-dimensional space with the assumption that no subsets of (n + 1) points lies on an (n − 1)-dimensional plane, we can use the recursion relation

$$D(N,n) = D(N - 1,n) + D(N - 1, n - 1) \tag{2.37}$$

to solve for $D(N,n)$. In particular,

$$D(1,n) = 2 \quad \text{and} \quad D(N,1) = 2N \tag{2.38}$$

Then

$$D(N,n) = \begin{cases} 2 \sum_{k=0}^{n} \binom{N - 1}{k} & N > n + 1 \\ \\ 2^N & N \leqslant n + 1 \end{cases} \tag{2.39}$$

where

$$\binom{N - 1}{k} = \frac{(N - 1)!}{k!(N - 1 - k)!}$$

Now, let us generalize this problem by finding the probability of the dichotomy that can be implemented. Given a ϕ machine and a set of N patterns in the pattern space, there are 2^N possible dichotomies and any one of the 2^N dichotomies can be picked up with probability

$$p = 2^{-N}$$

For the general decision function

$$d(\underline{x}) = \phi(\underline{x}) = w_1 f_1(\underline{x}) + w_2 f_2(\underline{x}) + \cdots + w_M f_M(\underline{x}) + w_{M+1} \quad (2.40)$$

the probability $p_{N,M}$ that any one dichotomy can be implemented is

$$p_{N,M} = \frac{\text{number of } \phi \text{ dichotomies}}{\text{total possible number of dichotomies}}$$

$$= \frac{D(N,M)}{2^N} \quad (2.41)$$

$$p_{N,M} = 2^{-N} \begin{cases} 2 \sum_{k=0}^{M} \binom{N-1}{k} & N > M + 1 \\[2em] 2^N & N \leqslant M + 1 \end{cases} \quad (2.42)$$

or

$$p_{N,M} = \begin{cases} 2^{1-N} \sum_{k=0}^{M} \binom{N-1}{k} & N > M + 1 \\[2em] 1 & N \leqslant M + 1 \end{cases} \quad (2.43)$$

Note that $p_{N,M} = 1$ for $N \leqslant M + 1$, which means that if the number of patterns is less than the number of adjustable weights for the general decision function, the patterns will always be linearly separable in the M-dimensional pattern space. But when $N > M + 1$, the probability of dichotomization will go down depending on the ratio of N to M + 1.

The analysis given above does not tell us how to choose $d(\underline{x})$ or $\phi(\underline{x})$, but it does tell us something about the ability of the machine to dichotomize patterns. For example, if we have a total of N patterns that properly represent two classes, we can be almost sure of finding a good classifier if M is large.

Example For a two-class three-dimensional pattern problem with a quadratic discriminant function $d(\underline{x})$, we have

$$M = \frac{n(n + 3)}{2} = 9$$

Then the capacity of dichotomization

$$C = 2(M + 1) = 20$$

If $N < 20$, we have a pretty good choice! This example also tells us how many prototype patterns we need for a good training set.

2.6 POTENTIAL FUNCTIONS AS DISCRIMINANT FUNCTIONS

A potential function $\psi(\underline{x}, \underline{z}_k^m)$ is known as a *kernel* in the probability density function estimator, or is a function of $\underline{x}$ and $\underline{z}_k^m$ defined over the pattern space, where $\underline{z}_k^m$ is the mth prototype defining class ω_k. The potential function is better illustrated by Fig. 2.15 for a one-dimensional pattern space. This potential gives the decreasing relationship between point $\underline{z}_k^m$ and point $\underline{x}$ as the distance $d(\underline{x}, \underline{z}_k^m)$ between these two points increases.

Superposition of the individual kernel "potential" functions will be used as a discriminant function

$$d_k(\underline{x}) = \frac{\sum\limits_{m=1}^{N_K} \psi(\underline{x}, \underline{z}_k^m)}{N_k} \tag{2.44}$$

which is defined for class k, where N_k is the number of prototypes in class k. Functions ψ may be different between classes or even between prototypes within a class. The average of these potentials of

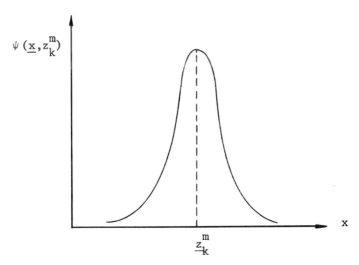

FIG. 2.15 Potential function of one variable.

prototypes from a given class indicates a degree of membership of the point $\underline{x}$ in the class.

The following characteristics of ψ are desirable:

1. $\psi(\underline{x},\underline{z})$ should be maximum for $\underline{x} = \underline{z}$.
2. $\psi(\underline{x},\underline{z})$ should be approximately zero for $\underline{x}$ distant from $\underline{z}$ in the region of interest.
3. $\psi(\underline{x},\underline{z})$ should be a smooth (continuous) function and tend to decrease in a monotonic fashion with the increase of the distance $d(\underline{x},\underline{z})$.
4. If $\psi(\underline{x}_1,\underline{z}) = \psi(\underline{x}_2,\underline{z})$, patterns $\underline{x}_1$ and $\underline{x}_2$ should have approximately the same degree of similarity to $\underline{z}$.

If a set of potential functions are found which form a satisfactory discriminant function as

$$d_k(\underline{x}) > d_j(\underline{x}) \quad \text{when } \underline{x} \in \omega_k \quad \forall \ j, k \qquad (2.45)$$

then

$$f(\underline{x}) + d_k(\underline{x}) > f(\underline{x}) + d_j(\underline{x}) \quad \forall \ f(\underline{x}) \qquad (2.46)$$

and

$$f(\underline{x})d_k(\underline{x}) > f(\underline{x})d_j(\underline{x}) \qquad \forall \ f(\underline{x}) > 0 \qquad (2.47)$$

This will help to simplify the computation of ψ and ultimately the computation of $d(\underline{x})$. Since, for example, if

$$\psi_1(\underline{x},\underline{z}) = \exp[-(\underline{x} - \underline{z})^T(\underline{x} - \underline{z})]$$

$$= \exp[-|\underline{x}|^2 - |\underline{z}|^2 + 2\underline{x}^T\underline{z}] \qquad (2.48)$$

after multiplying $\psi_1(\underline{x},\underline{z})$ by $f(\underline{x}) = \exp|\underline{x}|^2$, we obtain

$$\psi_2 = f(\underline{x})\psi_1(\underline{x},\underline{z})$$

$$= \exp[2\underline{x}^T\underline{z} - |\underline{z}|^2] \qquad (2.49)$$

which will be much simpler than that of $\psi_1(\underline{x},\underline{z})$.

Another form of potential function can also be chosen for sample pattern $\underline{z}$:

$$\psi(\underline{x},\underline{z}) = \sum_{i=1}^{r} \lambda_i^2 \phi_i(\underline{x})\phi_i(\underline{z}) \qquad (2.50)$$

where λ_i, $i = 1, 2, \ldots$ are constants and ϕ_i, $i = 1, 2, \ldots$, are orthonormal functions such that

$$\sum_{i=1}^{\infty} \phi_i(\underline{x})\phi_i(\underline{z}) = \delta(\underline{x} - \underline{z}) \qquad (2.51)$$

If $[\phi_i]$ is a complete orthonormal set, then for the decision function d_k,

$$d_k(\underline{x}) = \frac{1}{N_k} \sum_{m=1}^{N_k} \psi(\underline{x}, \underline{z}_k^m)$$

$$= \frac{1}{N_k} \sum_{m=1}^{N_k} \sum_{i=1}^{r} \lambda_i^2 \phi_i(\underline{x}) \phi_i(\underline{z}_k^m)$$

$$= \sum_{i=1}^{r} \phi_i(\underline{x}) \frac{1}{N_k} \sum_{m=1}^{N_k} \phi_i(\underline{z}_k^m)$$

$$= \sum_{i=1}^{r} c_i^k \phi_i(\underline{x}) \qquad\qquad (2.52)$$

where

$$c_i^k = \frac{1}{N_k} \sum_{m=1}^{N_k} \phi_i(\underline{z}_k^m)$$

This procedure is most attractive when either the number of samples N_k is small or the dimensionality of $\underline{x}$ is sufficiently small to allow $d(\underline{x})$ to be stored as a table for discrete values of $\underline{x}$. But if the number of samples is large, computation problems will be severe and storage problems may occur also.

PROBLEMS

2.1 Let $\underline{Y} = [\underline{y}_1, \underline{y}_2, \ldots, \underline{y}_M]$ be the set of all sample points. Find the normalized variables. Note: In the derivation of the expression for the normalized variables, N classes and M_N samples for the class N are assumed.

2.2 Discuss whether it would be successful for us to apply the method of successive dichotomies to the problem described by the following figure.

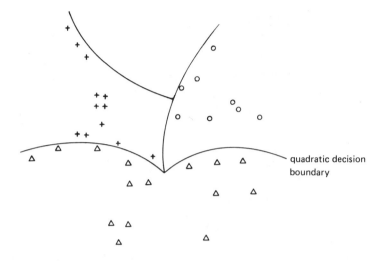

2.3 You are given the following prototypes in augmented space:

$$[(6,4,1), (5,2,1), (1,3,1), (0,-5,1)] \in S_1$$

$$[(5,-1,1), (4,-1,1), (3,1,1)] \in S_2$$

You are also informed that a layered machine (three TLUs in the first layer and one TLU in the second layer) might be a useful tool in dichotomizing the prototypes properly. Suppose that the following weight vectors have been selected for the first-layer TLU:

$$\underline{W}_1 = -1, 1, 4$$

$$\underline{W}_2 = 1, 1, -1$$

$$\underline{W}_3 = -\frac{1}{2}, 1, 0$$

(a) Compute and plot the prototypes in the first layer space.

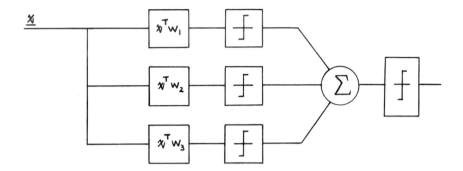

(b) Will a committee machine separate these prototypes with the TLUs as shown?

(c) Which prototype must be removed for the first-layer space to be linearly separable?

(d) What happens to the committee machine when the weight vectors are changed to the following?

$$\underline{W}_1 = -1, \ -1, \ 4$$

$$\underline{W}_2 = 1, \ -1, \ -1$$

$$\underline{W}_3 = -\frac{1}{2}, \ -1, \ 0$$

2.4 The following three decision functions are given for a three-class problem:

$$d_1(\underline{x}) = 10x_1 - x_2 - 10 = 0$$

$$d_2(\underline{x}) = x_1 + 2x_2 - 10 = 0$$

$$d_3(\underline{x}) = x_1 - 2x_2 - 10 = 0$$

(a) Sketch the decision boundary and regions for each pattern class, assuming that each pattern class is separable from all other classes by a single decision surface.

(b) Assuming that each pattern class is pairwise separable from every other class by a distinct decision surface, and letting $d_{12}(\underline{x}) = d_1(\underline{x})$, $d_{13}(\underline{x}) = d_2(\underline{x})$, and $d_{23}(\underline{x}) = d_3(\underline{x})$ as listed above, sketch the decision boundary and regions for each pattern class.

3
Nonparametric (Distribution-Free) Training of Discriminant Functions

3.1 WEIGHT SPACE

We have already discussed the fact that a pattern vector $\underline{x}$ appears as a point in the pattern space, and that a pattern space can be partitioned into subregions for patterns belonging to different categories. The decision surfaces that partition the space may be linear, piecewise linear, or nonlinear, and can be generalized as

$$d(\underline{x}) = f(\underline{w}, \underline{x}) \tag{3.1}$$

where

$$\underline{x} = (x_1, x_2, \ldots, x_n, 1)^T \quad \text{and} \quad \underline{w} = (w_1, w_2, \ldots, w_n, w_{n+1})^T$$

represent the pattern and weight vectors, respectively. The problem of training a system is actually to find the weight vector $\underline{w}$ shown in Eq. (3.1) with the a priori information obtained from the training samples. It is possible and perhaps more convenient to investigate the behavior of the training algorithms in a weight space. The weight space is an $(n + 1)$-dimensional euclidean space in which the coordinate variables are w_1, w_2, $\ldots$, w_n, w_{n+1}. For each prototype $\underline{z}_k^m$, $k = 1$, 2, $\ldots$, M, m = 1, 2, $\ldots$, N_k (where M represents the number of categories and N_k represents the number of prototypes belonging to category k), there is in W space (weight space) a hyperplane on which

$$\underline{w}^T \underline{z}_k^m = 0 \tag{3.2}$$

Any weight vector $\underline{w}$ on the positive side of the hyperplane yields $\underline{w}^T \underline{z} > 0$. That is, if the prototype $\underline{z}_k^m$ belongs to category ω_1, any weight vector $\underline{w}$ on this side of the hyperplane will probably classify $\underline{z}_k^m$. A similar argument can be made for any weight vector on the other side of this hyperplane, where $\underline{w}^T \underline{z} < 0$.

Let us take a two-class problem for illustration. Suppose that we have a set of N_1 patterns belonging to ω_1 and a set of N_2 patterns belonging to ω_2, with the total number of patterns $N = N_1 + N_2$. Assume also that ω_1 and ω_2 are two linearly separable classes. Then a vector $\underline{w}$ can be found such that

$$\underline{w}\,\underline{z}_1^m > 0 \quad \forall \quad \underline{z}_1^m \in \omega_1, \quad m = 1, 2, \ldots, N_1 \qquad (3.3)$$

and

$$\underline{w}\,\underline{z}_2^m < 0 \quad \forall \quad \underline{z}_2^m \in \omega_2, \quad m = 1, 2, \ldots, N_2 \qquad (3.4)$$

where $\underline{z}_1^m$ and $\underline{z}_2^m$ represent all the prototypes in categories ω_1 and ω_2, respectively. In general, for N patterns there are N pattern hyperplanes in the weight space. The solution region for category ω_1 in W space is that region which lies on the positive side of the N_1 hyperplanes for category ω_1 and on the negative side of the N_2 hyperplanes for category ω_2.

Suppose that we have three prototypes $\underline{z}^1$, $\underline{z}^2$, and $\underline{z}^3$, and know that all of them belong to category ω_1. Three hyperplanes can then be drawn in the W space, as shown in Fig. 3.1a. The shaded area in Fig. 3.1a shows the solution region in this two-class problem. In this region

$$d_1(\underline{w},\underline{z}) > 0$$

$$d_2(\underline{w},\underline{z}) > 0$$

and

$$d_3(\underline{w},\underline{z}) > 0$$

That is, any $\underline{w}$ in this region will probably classify the prototypes $\underline{z}^1$, $\underline{z}^2$, and $\underline{z}^3$ as belonging to category ω_1, while in the crosshatched area shown in Fig. 3.1b,

$$d_1(\underline{w},\underline{z}) > 0$$

$$d_2(\underline{w},\underline{z}) > 0$$

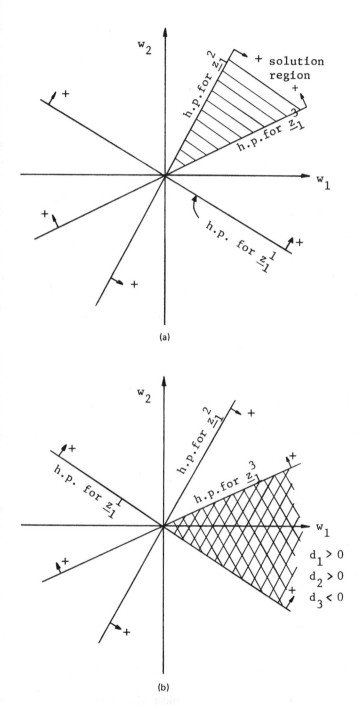

FIG. 3.1 Hyperplanes in W space.

but

$$d_3(\underline{w}, \underline{z}) < 0$$

Any $\underline{w}$ over this region will classify $\underline{z}^1$ and $\underline{z}^2$ as being in category ω_1, and classify $\underline{z}^3$ as being in category ω_2.

As discussed in Chap. 2, the decision surface for a two-class problem is assumed to have the property that $d(\underline{w},\underline{x})$ will be greater than zero for all patterns of one class, but less than zero for all patterns belonging to the other class. But if all the $\underline{z}_2^m$'s are replaced by their negatives, $-\underline{z}_2^m$'s, the solution region can be generalized as that part of the $\underline{W}$ space for which

$$\underline{w}^T \underline{z} > 0 \quad \forall \quad \underline{z} = \underline{z}_1^m, \ -\underline{z}_2^m \tag{3.5}$$

Our problem then simply becomes to find a $\underline{w}$ such that all inequalities are greater than zero.

It might be desirable to have a margin (or threshold) in the discriminant function such that

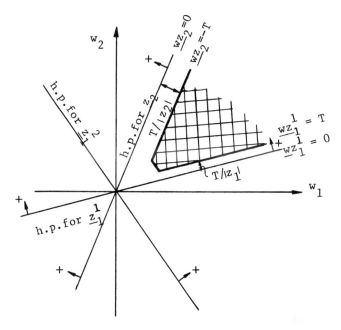

FIG. 3.2 Solution region for a two-class problem with margin set for each decision surface.

$$\underline{w}^T \underline{z} > T \quad \forall \quad \underline{z} = \underline{z}_1^m, \; -\underline{z}_2^m \tag{3.6}$$

where $T > 0$ is the margin (or threshold) chosen. Any $\underline{w}$ satisfying inequality (3.6) is a weight solution vector. The solution region is now changed as shown in Fig. 3.2. In the crosshatched region, both $\underline{w}^T \underline{z}_1^1$ and $\underline{w}^T \underline{z}_1^2$ are positive, while $\underline{w}^T \underline{z}_2 < 0$. Note that along the original pattern hyperplane

$$\underline{w}^T \underline{z} = \underline{w} \cdot \underline{z} = 0 \tag{3.7}$$

and that the vector $\underline{z}$ (augmented $\underline{z}$) is perpendicular to the hyperplane $\underline{w}^T \underline{z} = 0$ and heads in its positive direction. Thus the line $\underline{w}^T \underline{z} = T$ is offset from $\underline{w}^T \underline{z} = 0$ by a distance $\Delta = T / |\underline{z}|$. The proof of this is left to the reader as a problem.

3.2 ERROR CORRECTION TRAINING PROCEDURES

It is obvious that for a two-class problem an error would exist if

$$\underline{w}^T \underline{z}_1^m < 0(T) \tag{3.8}$$

$$\underline{w}^T \underline{z}_2^m > 0(T) \tag{3.9}$$

Then we need to move the weight vector $\underline{w}$ to the positive side of the pattern hyperplane for $\underline{z}_1^m$, in other words, move the vector $\underline{w}$ to the correct solution region.

The most direct way of doing this is to move $\underline{w}$ in a direction perpendicular to the hyperplane (i.e., in a direction of $\underline{z}_1^m$ or $-\underline{z}_2^m$). In general, the correction of the $\underline{w}$ can be formulated as: Replace $\underline{w}(k)$ by $\underline{w}(k + 1)$ such that

$$\underline{w}(k + 1) = \underline{w}(k) + c\underline{z}_1^m \quad \text{if } \underline{w}^T(k)\underline{z}_1^m < 0(T)$$

$$\underline{w}(k + 1) = \underline{w}(k) - c\underline{z}_2^m \quad \text{if } \underline{w}^T(k)\underline{z}_2^m > 0(-T) \tag{3.10}$$

$$\underline{w}(k + 1) = \underline{w}(k) \quad \text{if correctly classified}$$

where $\underline{w}(k)$ and $\underline{w}(k + 1)$ are the weight vectors at the kth and $(k + 1)$th correction steps, respectively. To add a correction term $c\underline{z}_1^m$ implies moving vector $\underline{w}$ in the direction of $\underline{z}_1^m$. Similarly, subtracting a correction term $c\underline{z}_2^m$ implies moving vector $\underline{w}$ in the direction of $-\underline{z}_2^m$.

During this training period, patterns are presented one at a time through all $N = N_1 + N_2$ prototypes (training patterns). Each complete pass through all the N patterns is called an *iteration*. After one iteration, all patterns are presented again in the same sequence to carry on another iteration. This is repeated until no corrections are made through one complete iteration.

Several rules can be set up in choosing the value of c: the fixed increment rule, absolute correction rule, fractional correction rule, and so on.

3.2.1 Fixed-Increment Rule

In this algorithm, c is chosen to be a positive fixed constant. This algorithm begins with any $\underline{w}(0)$, and Eq. (3.10) is applied to the training sequence P, where $P = [\underline{z}_1^1, \underline{z}_2^1, \ldots, \underline{z}_1^{N_1}, \underline{z}_2^{N_2}]$. The whole weight-adjustment process will terminate in some finite steps, k.

The choice of c for this process is actually not important. If the theorem holds for c = 1, it holds for any c ≠ 1, since this, in effect, just scales all patterns by some amount without changing their separability.

3.2.2 Absolute Correction Rule

In this algorithm c is chosen to be the smallest integer that will make $\underline{w}(k + 1)$ cross the pattern hyperplane into the solution region of W space each time a classification error is made. Let $\underline{z}_i'$ be the average of the sample vectors that do not satisfy the inequality $\underline{w} \cdot \underline{z} \geqslant T$. The constant c is chosen such that

$$\underline{w}^T(k + 1)\underline{z}_i' = [\underline{w}(k) + c\underline{z}_i']^T \underline{z}_i' > T \qquad (3.11)$$

or

$$c\underline{z}_i'^T \underline{z}_i' > T - \underline{w}^T(k)\underline{z}_i' > 0 \qquad (3.12)$$

and therefore

$$c > \frac{T - \underline{w}^T(k)\underline{z}_i'}{\underline{z}_i'^T \underline{z}_i'} = \frac{T - \underline{w}^T(k)\underline{z}_i'}{|\underline{z}_i'|^2} \qquad (3.13)$$

Note that if T = 0, $-\underline{w}^T(k)\underline{z}_i'$ must be greater than zero, or $\underline{w}^T(k)\underline{z}_i' < 0$. Taking its absolute value into consideration, Eq. (3.13) becomes

$$c > \frac{|\underline{w}^T(k)\underline{z}_i'|}{|\underline{z}_i'|^2} \tag{3.14}$$

The absolute correction rule will also yield a solution weight vector in a finite number of steps.

3.2.3 Fractional Correction Rule

In W space, the augmented pattern vector $\underline{z}$ is perpendicular to the hyperplane $\underline{z}^T\underline{w} = 0$ and heads in the positive direction, as shown in Fig. 3.3. The distance from $\underline{w}(k)$ to the desired hyperplane is

$$D = \Delta + \delta = \frac{T}{|\underline{z}_i'|} + \left| \underline{w}(k) \cdot \frac{\underline{z}_i'}{|\underline{z}_i'|} \right|$$

$$= \frac{T}{|\underline{z}_i'|} + \frac{|\underline{w}^T(k)\underline{z}_i'|}{|\underline{z}_i'|} \tag{3.15}$$

When $\underline{w}(k)$ is on the other side of the hyperplane,

$$D = \frac{T - |\underline{w}^T(k)\underline{z}_i'|}{|\underline{z}_i'|} \tag{3.16}$$

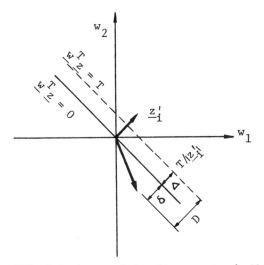

FIG. 3.3 Augmented pattern vector in W space.

In the fractional correction algorithm c is chosen so that $\underline{w}$ is moved by a fraction of the distance in a direction normal to the desired hyperplane. That is,

$$c = \lambda \, \frac{D}{|\underline{z}'_i|} \qquad \lambda > 0 \tag{3.17}$$

and

$$\underline{w}(k + 1) - \underline{w}(k) = \lambda D \, \frac{\underline{z}'_i}{|\underline{z}'_i|} \tag{3.18}$$

If the threshold is set at 0, then

$$c = \lambda \, \frac{|\underline{w}^T(k)\underline{z}'_i|}{|\underline{z}'_i|^2} \qquad \lambda > 0 \tag{3.19}$$

It can be seen that when $\lambda = 1$, the correction is to the hyperplane (absolute correction rule); when $\lambda < 1$, the correction is short of the hyperplane (underrelaxation); and when $\lambda > 1$, the correction is over the hyperplane (overrelaxation). For $0 < \lambda < 2$, the fractional correction rule will either terminate on a solution weight vector in a finite number of steps or else converge to a point on the boundary of the solution weight space.

The training procedure can then be generalized for any of the foregoing three algorithms as follows:

1. Take each $\underline{z}$ from the training set and test $d(\underline{z})$ for its category belonging. $M = 2$ is assumed here.
2. If a correct answer is obtained, go to the next $\underline{z}$.
3. If misclassification occurs, correct $\underline{w}(k)$ with $\underline{w}(k + 1)$.
4. After all $\underline{z}$'s from the training set have been examined, repeat the entire process in the same sequential order. If the $\underline{z}$'s are linearly separable, all three of these algorithms will converge to a correct $\underline{w}$.

Figure 3.4 shows the correction steps for the three different procedures. Absolute correction terminates in three steps, whereas fractional correction terminates in four steps.

For classes greater than two ($M > 2$), similar procedures can be followed. Assume that we have training sets available for all pattern classes ω_i, $i = 1, 2, \ldots, M$. Compute the discriminant function $d_i(\underline{z}) = \underline{w}_i\underline{z}$, $i = 1, 2, \ldots, M$. Obviously, we desire

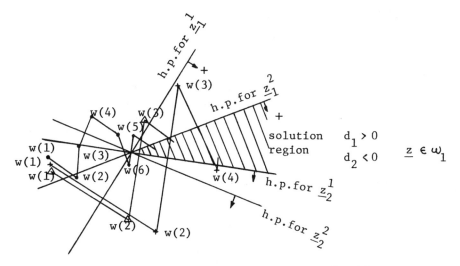

FIG. 3.4 Weight vector adjustment procedures. ●, fixed increment; △, absolute correction; +, fractional correction.

$$d_i(\underline{z}) > d_j(\underline{z}) \quad \text{if} \quad \underline{z} \in \omega_i \quad \forall \; j \neq i \qquad (3.20)$$

If so, the weight vectors are not adjusted. But if

$$d_j(\underline{z}) > d_i(\underline{z}) \quad \text{if} \quad \underline{z} \in \omega_i \quad \forall \; j \neq i \qquad (3.21)$$

misclassification occurs, and weight adjustment will be needed. Under these circumstances the following adjustment can be made for the fixed-increment correction rule:

$$\underline{w}_i(k + 1) = \underline{w}_i(k) + c\underline{z}(k) \qquad (3.22)$$

$$\underline{w}_j(k + 1) = \underline{w}_j(k) - c\underline{z}(k) \qquad (3.23)$$

$$\underline{w}_\ell(k + 1) = \underline{w}_\ell(k) \qquad (3.24)$$

where k and k + 1 denote the kth and (k + 1)th iteration steps. Equation (3.23) is for those j's that make $d_j(\underline{z}) > d_i(\underline{z})$, and Eq. (3.24) is for those ℓ's that are neither i nor those making an incorrect classsification. If the classes are separable, this algorithm will converge in a finite number of iterations. Similar adjustments can be derived for the absolute and fractional correction algorithms.

3.3 GRADIENT TECHNIQUES

3.3.1 General Gradient Descent Technique

The gradient descent technique is another approach to training the system. A gradient vector possesses the important property of pointing in the direction of the maximum rate of increase of the function when the argument increases. The weight-adjustment procedure can then be formulated as

$$\underline{w}(k + 1) = \underline{w}(k) - \rho_k \nabla J(\underline{w}) \Big|_{w=\underline{w}(k)} \tag{3.25}$$

where $J(\underline{w})$ is an index of performance or a criterion function that is to be minimized by adjusting $\underline{w}$. Minimum $J(\underline{w})$ can be approached by moving $\underline{w}$ in the direction of the negative of the gradient. The procedure can be summarized as follows:

1. Start with some arbitrarily chosen weight vector $\underline{w}(1)$ and compute the gradient vector $\nabla J[\underline{w}(1)]$.
2. Obtain the next value $\underline{w}(2)$ by moving some distance from $\underline{w}(1)$ in the direction of steepest descent.

ρ_k in Eq. (3.25) is a positive scale factor that sets the step size. For its optimum choice, let us assume that $J(\underline{w})$ can be approximated by

$$J[\underline{w}(k + 1)] \simeq J[\underline{w}(k)] + [\underline{w}(k + 1) - \underline{w}(k)]^T \nabla J[\underline{w}(k)]$$
$$+ \frac{1}{2} [\underline{w}(k + 1) - \underline{w}(k)]^T D[\underline{w}(k + 1) - \underline{w}(k)]$$

$$\tag{3.26}$$

where

$$D = \frac{\partial^2 J}{\partial w_i \, \partial w_j} \Bigg|_{w=\underline{w}(k)}$$

Substitution of Eq. (3.25) into Eq. (3.26) yields

$$J[\underline{w}(k + 1)] \simeq J[\underline{w}(k)] - \rho_k \big| \nabla J[\underline{w}(k)] \big|^2 + \frac{1}{2} \rho_k^2 \nabla J^T D \nabla J \tag{3.27}$$

Setting $\partial J[w(k + 1)]/\partial \rho_k = 0$ for minimum $J[\underline{w}(k + 1)]$, we obtain

$$\big| \nabla J \big|^2 = \rho_k \nabla J^T D \nabla J \tag{3.28}$$

or

$$\rho_k = \left. \frac{|\nabla J|^2}{\nabla J^T D \, \nabla J} \right|_{\underline{w} = \underline{w}(k)} \tag{3.29}$$

which is equivalent to Newton's algorithm for optimum descent, in which

$$\rho_k = D^{-1} \tag{3.30}$$

Some problems may exist with this optimum ρ_k: D^{-1} in Eq. (3.29) may not exist; the matrix operations involved are time consuming and expensive; and the assumption of the second-order surface may be incorrect. For those reasons, setting ρ_k equal to a constant may do just as well.

3.3.2 Perceptron Criterion Function

Let the criterion function be

$$J_p(\underline{w}) = \sum_{\underline{z} \in P} -\underline{w}^T \underline{z} \tag{3.31}$$

where the summation is over the incorrectly classified pattern samples. Geometrically, $J_p(\underline{w})$ is proportional to the sum of the distances of the misclassified patterns to the hyperplane. Taking the derivative of $J_p(\underline{w})$ with respect to $\underline{w}(k)$ yields

$$\nabla J_p[\underline{w}(k)] = - \sum_{\underline{z} \in P} \underline{z} \tag{3.32}$$

where $\underline{w}(k)$ denotes the value of $\underline{w}$ at the kth iteration step. The perceptron training algorithm can then be formulated as

$$\underline{w}(k + 1) = \underline{w}(k) - \rho_k \nabla J_p[\underline{w}(k)] \tag{3.33}$$

or

$$\underline{w}(k + 1) = \underline{w}(k) + \rho_k \sum_{\underline{z} \in P} \underline{z} \tag{3.34}$$

where P is the set of samples misclassified by $\underline{w}(k)$. Equation (3.34) can be thus interpreted to mean that the (k + 1)th weight vector can be obtained by adding some multiple of the sum of the misclassified samples to the kth weight vector. This is a "many-at-a-time" procedure, since we determine $\underline{w}^T \underline{z}$ for all $\underline{z} \in P$ and adjust only after all patterns were classified.

If we make an adjustment after each incorrectly classified pattern (we call it "one-at-a-time" procedure), the criterion function becomes

$$J(\underline{w}) = -\underline{w}^T \underline{z} \tag{3.35}$$

and

$$\nabla J(\underline{w}) = -\underline{z} \tag{3.36}$$

The training algorithm is to make

$$\underline{w}(k + 1) = \underline{w}(k) + \rho_k \underline{z} \tag{3.37}$$

This is the fixed-increment rule if $\rho_k = c$ (a constant).

3.3.3 Relaxation Criterion Function

The criterion function used in this algorithm is chosen to be

$$J_r(\underline{w}) = \frac{1}{2} \sum_{z \in P} \frac{(-\underline{w}^T \underline{z} + \underline{b})^2}{|\underline{z}|^2} \tag{3.38}$$

Again, P is the set of samples misclassified by $\underline{w}$. That is, P consists of those $\underline{z}$'s for which $-\underline{w}^T \underline{z} + \underline{b} > 0$ or $\underline{w}^T \underline{z} < \underline{b}$. The gradient of $J_r(\underline{w})$ with respect to $\underline{w}(k)$ yields

$$\nabla J_r(\underline{w}) = - \sum_{\underline{z} \in P} \frac{-\underline{w}^T \underline{z} + \underline{b}}{|\underline{z}|^2} \underline{z} \tag{3.39}$$

The basic relaxation training algorithm is then formulated as

$$\underline{w}(k + 1) = \underline{w}(k) + \rho_k \sum_{\underline{z} \in P} \frac{-\underline{w}^T(k)\underline{z} + \underline{b}}{|\underline{z}|^2} \underline{z} \tag{3.40}$$

This is also a many-at-a-time algorithm. Its corresponding one-at-a-time algorithm is

$$w(k + 1) = w(k) + \rho_k \frac{-w^T(k)z + b}{|z|^2} z \qquad (3.41)$$

which becomes the fractional correction algorithm with $\lambda = \rho_k$.

3.4 TRAINING OF PIECEWISE LINEAR MACHINES

In general, no convergence theorems exist for the training procedures of committee (or other piecewise linear) machines. One procedure that frequently is satisfactory is given here. Assume that M = 2 and that there are R discriminant functions, where R is odd. Then

$$d_i(z) = w_i^T(k)z \qquad i = 1, 2, \ldots, R \qquad (3.42)$$

The classification of the committee machines will then be made according to

$$d(z) = \sum_{i=1}^{R} \text{sgn } d_i(z) \qquad (3.43)$$

such that

z is assigned to ω_1 when $d(z) > 0$

z is assigned to ω_2 when $d(z) < 0$ \qquad (3.44)

where

$$\text{sgn } d_i(z) = \begin{cases} +1 & \text{if } d_i(z) \geqslant 0 \\ -1 & \text{if } d_i(z) < 0 \end{cases} \qquad (3.45)$$

Note that since R is odd, $d(z)$ cannot be zero and will also be odd. Thus $d(z)$ is equal to the difference between the number of $d_i(z) > 0$ and that of $d_i(z) < 0$ for a weight vector $w_i(k)$ at the kth iteration step. In this regard, we always desire $d(z) > 0$. In other words, we want to have more weight vectors that yield $d_i(z) > 0$ than those which yield $d_i(z) < 0$.

When $d(z) < 0$, incorrect classification results. It will be obvious that in this case there will be $[R + d(z)]/2$ weight vectors among the $w_i(k)$, $i = 1, \ldots, R$, which yield negative responses $[d_i(z) < 0]$

and $[R - d(\underline{z})]/2$ weight vectors which yield positive responses $[d_i(\underline{z}) > 0]$. To obtain correct classification, we then need to change at least n responses of the $w_i(k)$ from -1 to $+1$, where n can be found by setting up Eq. (3.46)

$$\left[\frac{R - d(\underline{z})}{2} + n \right] - \left[\frac{R + d(\underline{z})}{2} - n \right] = 1 \qquad (3.46)$$

The first set of brackets represents the number of d_i that are presently greater than zero; the set of brackets after the minus sign represents the number of d_i that are presently less than zero. The minimum value of n is then

$$n_{min} = \frac{d(\underline{z}) + 1}{2} \qquad (3.47)$$

which is the minimum number of weight vectors needed to be adjusted. The procedure for the weight vector adjustment will then be as follows:

1. Pick out the least negative $d_i(\underline{z})$'s among those negative $d_i(\underline{z})$'s.
2. Adjust the $[d(\underline{z}) + 1]/2$ weight vectors that have the least negative $d_i(\underline{z})$'s by the following rule:

$$\underline{w}_i(k + 1) = \underline{w}_i(k) + c\underline{z} \qquad (3.48)$$

so that their resulting $d_i(\underline{z})$'s become positive. All the other weight vectors are left unaltered at this stage.
3. If at the kth stage, the machine incorrectly classifies a pattern that should belong to ω_2, give the correction increment c a negative value, such as

$$\underline{w}_i(k + 1) = \underline{w}_i(k) - c\underline{z} \qquad (3.49)$$

3.5 PRACTICAL CONSIDERATIONS CONCERNING ERROR CORRECTION TRAINING METHODS

Since error-correcting rules never allow an error in pattern classification without adjusting the discriminant function, some oscillations may result. For example, for the case of two normally distributed classes with overlap, an error will always occur even if the optimum hyperplane is found. The error correction rule will cause the hyperplane continually to be adjusted and never stabilize at the optimum location.

For the case that classes have more than one "cluster" or "grouping" in the pattern space, the error correction training method will again encounter problems. The remedy is to add a stopping rule. But this stopping rule must be employed appropriately; otherwise, the system may terminate on a poor $\underline{w}$. Another way of solving such problems is to go to a training procedure that is not error correcting, such as clustering (determining only the modes of a multimodal problem), stochastic approximation, potential functions, or the minimum squared error procedure.

3.6 MINIMUM SQUARED ERROR PROCEDURES

3.6.1 Minimum Squared Error and the Pseudoinverse Method

Consider that we wish to have the equalities

$$\underline{z}\underline{w} = \underline{b} \qquad (3.50)$$

instead of the inequalities $\underline{z}\underline{w} > 0$. Then we are required to solve $N = \sum_{i=1}^{M} N_i$ linear equations, where N is the total number of prototypes for all classes, N_i is that for class ω_i, and M is the total number of classes. $\underline{z}$ and $\underline{b}$ can then be defined, respectively, as

$$
\underline{Z} =
\begin{vmatrix}
\underline{z}_1 \\
\underline{z}_2 \\
. \\
. \\
. \\
\underline{z}_n
\end{vmatrix}
=
\begin{vmatrix}
z_{11} & z_{12} & \cdots & z_{1n} \\
z_{21} & z_{22} & \cdots & z_{2n} \\
. & & & . \\
. & & & . \\
. & & & . \\
z_{N1} & z_{N2} & \cdots & z_{Nn}
\end{vmatrix}
\qquad (3.51)
$$

and

$$
\underline{b} =
\begin{vmatrix}
b_1 \\
b_2 \\
. \\
. \\
. \\
b_N
\end{vmatrix}
\qquad (3.52)
$$

If $\underline{z}$ were square and nonsingular, we could set

$$\underline{w} = \underline{z}^{-1}\underline{b} \qquad (3.53)$$

and solve for $\underline{w}$. But, in general, $\underline{z}$ is rectangular (i.e., more rows than columns) and many solutions to $\underline{z}\underline{w} = b$ exist. Let us define an error vector as

$$\underline{e} = \underline{z}\underline{w} - \underline{b} \qquad (3.54)$$

and a sum-of-squared-error criterion function as

$$J_s(\underline{w}) = \frac{1}{2}|\underline{e}|^2 = \frac{1}{2}|\underline{z}\underline{w} - \underline{b}|^2 = \frac{1}{2}\sum_{i=1}^{N} (\underline{z}_i\underline{w} - \underline{b}_i)^2 \qquad (3.55)$$

Taking the partial derivative of J_s with respect to $\underline{w}$, we obtain

$$\nabla J_s(\underline{w}) = \sum_{i=1}^{N} (\underline{z}_i\underline{w} - \underline{b}_i)\underline{z}_i \qquad (3.56)$$

or in matrix form,

$$\nabla J_s(\underline{w}) = \underline{z}^T(\underline{z}\,\underline{w} - \underline{b}) \qquad (3.57)$$

Setting $\nabla J_s(\underline{w}) = 0$ yields

$$\underline{z}^T\underline{w}\underline{z} = \underline{z}^T\underline{b} \qquad (3.58)$$

or

$$\underline{w} = \underline{z}^{\#}\underline{b} \qquad (3.59)$$

where $z^{\#} = (\underline{z}^T\underline{z})^{-1}\underline{z}^T$ is called the *pseudoinverse* or generalized inverse of $\underline{z}$. $\underline{z}^{\#}$ has the following properties: (1) $\underline{z}^{\#}\underline{z} = I$, but in general, $\underline{z}\underline{z}^{\#} \neq I$; and (2) $\underline{z}^{\#} = \underline{z}^{-1}$ if $\underline{z}$ is square and nonsingular. The value of b in Eq. (3.52) may be set arbitrarily except that $b_i > 0 \; \forall \; i$. If no other information is available, a good choice is

$$\underline{b} = [1 \quad 1 \quad 1 \quad \cdots \quad 1] = \underline{u}^T$$

In fact, if $\underline{b} = \underline{u}^T$, the minimum squared error solution approaches a minimum mean squared error approximation to the Bayes discriminant function. Note that this method is not error correcting, since it does not compute new $\underline{w}$ for every $\underline{z}$. In fact, all $\underline{z}$ are considered together and only one solution is needed; therefore, the training time is very short.

3.6.2 Ho-Kashyap Method

When the criterion function $J(\underline{w})$ is to be minimized not only with respect to $\underline{w}$, but also with respect to $\underline{b}$ (i.e., assume that $\underline{b}$ is not a constant), the training algorithm is known as the Ho-Kashyap method. The same criterion function J as that shown in Eq. (3.55) is to be used and repeated here:

$$J(\underline{w}) = \frac{1}{2} \left| \underline{z}\,\underline{w} - \underline{b} \right|^2 \tag{3.60}$$

Partial derivatives of $J(\underline{w})$ with respect to $\underline{w}$ and $\underline{b}$ are, respectively,

$$\frac{\partial J(\underline{w})}{\partial w} = \underline{z}^T (\underline{z}\,\underline{w} - \underline{b}) \tag{3.61}$$

and

$$\frac{\partial J(\underline{w})}{\partial b} = -(\underline{z}\,\underline{w} - \underline{b}) \tag{3.62}$$

Setting $\partial J / \partial w = 0$ yields

$$\underline{w} = (\underline{z}^T \underline{z})^{-1} \underline{z}^T \underline{b} = \underline{z}^{\#} \underline{b} \tag{3.63}$$

Since all components of $\underline{b}$ are constrained to be positive, adjustments on $\underline{b}$ can be made such that

$$\underline{b}(k + 1) = \underline{b}(k) + \delta\underline{b}(k) \tag{3.64}$$

where

$$\delta b_i(k) = \begin{cases} 2c[\underline{e}(k) & \text{when } \underline{e}(k) > 0 \\ \\ 0 & \text{when } \underline{e}(k) \leqslant 0 \end{cases} \tag{3.65}$$

where k, i, and c represent the iteration index, the component index of the vector, and the positive correction increment, respectively. From Eq. (3.63) we have

$$\underline{w}(k + 1) = \underline{z}^{\#}\underline{b}(k + 1) \tag{3.66}$$

Combining Eqs. (3.63), (3.64) and (3.66), we obtain

$$\underline{w}(k + 1) = \underline{w}(k) + \underline{z}^{\#}\delta\underline{b}(k) \tag{3.67}$$

Remembering that the components of $\underline{b} = (b_1, b_2, \ldots, b_N)^T$ are all positive, that is,

$$\underline{w}(1) = \underline{z}^{\#}\underline{b}(1) \qquad \underline{b}(1) > 0 \tag{3.68}$$

$$\underline{e}(k) = \underline{z}\,\underline{w}(k) - \underline{b}(k) \tag{3.69}$$

the algorithm for the weight and b adjustments can be put in the following form:

$$\underline{w}(k + 1) = \underline{w}(k) + c\underline{z}^{\#}[\underline{e}(k) + |\underline{e}(k)|] \tag{3.70}$$

$$\underline{b}(k + 1) = \underline{b}(k) + c[\underline{e}(k) + |\underline{e}(k)|] \tag{3.71}$$

3.6.3 Widrow-Hoff Rule

If either $\underline{z}^T\underline{z}$ is singular or the matrix operations in finding $\underline{z}^{\#}$ are unwieldy, we can minimize $J_s(\underline{w})$ by a gradient descent procedure:

Step 1: Choose $\underline{w}(0)$ arbitrarily.

Step 2: Adjust the weight vector such that

$$\underline{w}(k + 1) = \underline{w}(k) - \rho_k \nabla J(\underline{w}) \big|_{\underline{w}=\underline{w}(k)}$$

or

$$\underline{w}(k + 1) = \underline{w}(k) - \rho_k \underline{z}^T[\underline{w}(k)\underline{z} - \underline{b}]$$

If ρ_k is chosen to be ρ_1/k, it can be shown that this converges to a limiting weight vector $\underline{w}$ satisfying

$$\nabla J_s(\underline{w}) = \underline{z}^T(\underline{w}\underline{z} - \underline{b}) = 0$$

In this algorithm matrix operations are still required, but the storage requirements are usually less here than with the $\underline{z}^{\#}$ above.

PROBLEMS

3.1 Given the sample vectors

$$\underline{z}_1 = (0,0)$$

$$\underline{z}_2 = (-1,-1)$$

$$\underline{z}_3 = (2,2)$$

$$\underline{z}_4 = (4,0)$$

$$\underline{z}_5 = (4,1)$$

where $[\underline{z}_3,\underline{z}_4,\underline{z}_5] \in \omega_2$ and $[\underline{z}_1,\underline{z}_2] \in \omega_1$. If they are presented in numerical order repeatedly, give the sequence of weight vectors and the solution generated by using fixed increment correction rule. Start with $W^T(1) = 0$.

3.2 Repeat Problem 3.1 with the following sample vectors:

$$\underline{z}_1 = (0,0) \in \omega_1$$

$$\underline{z}_2 = (3,3) \in \omega_2$$

$$\underline{z}_3 = (-3,3) \in \omega_3$$

Start with $\underline{W}_1(1) = \underline{W}_2(1) = \underline{W}_3(1) = (0,0,0)^T$.

3.3 Given the following set of data:

$$\underline{z}_1 = (0,0,0)^T, \quad \underline{z}_2 = (1,0,0)^T, \quad \underline{z}_3 = (1,0,1)^T,$$

$$\underline{z}_4 = (1,1,0)^T \in \omega_1$$

$$\underline{z}_5 = (0,0,1)^T, \quad \underline{z}_6 = (0,1,1)^T, \quad \underline{z}_7 = (0,1,0)^T,$$

$$\underline{z}_8 = (1,1,1)^T \in \omega_2$$

Find a solution weight vector using the perceptron algorithm. Start with $\underline{W}(1) = (-1,-2,-2,0)^T$.

4

Statistical Discriminant Functions

In this chapter we discuss primarily statistical discriminant functions used to deal with those sorts of classification in which pattern categories are known a priori to be characterizable by a set of parameters. First, formulation of the classification problem by means of statistical decision theory is introduced, and loss functions, Bayes' discriminant function, maximum likelihood decision, and so on, are discussed. Some analysis of the probability error is given.

The optimal discriminant function for normally distributed patterns is then discussed in more detail, followed by a discussion of how to determine the probability density function when it is unknown. At the end of the chapter, a large-data-set aerial-photo interpretation problem is taken as an example to link the theory we have discussed with the real-world problem we actually have.

4.1 INTRODUCTION

The use of statistical discriminant functions for classification is advantageous because (1) considerable knowledge already exists in areas such as statistical communication, detection theory, decision theory, and so on, and this knowledge is directly applicable to pattern recognition; and (2) statistical formulation is particularly suitable for the pattern recognition problem, since many pattern recognition processes are modeled statistically. In pattern recognition it is desirable to use all the a priori information available and the performance of the system is also often evaluated statistically.

In the training of a statistical classification system, an underlying distribution density function such as gaussian distribution or some other distribution function is assumed; however, no known distribution is assumed in nonparametric training, as we discussed in Chaps. 2 and 3.

4.2 PROBLEM FORMULATION BY MEANS OF STATISTICAL DECISION THEORY

4.2.1 Loss Function

Before we establish the loss functions, it will be helpful to make the following assumptions:

1. $p(\omega_i)$ is known or can be estimated.
2. $p(\underline{x}|\omega_i)$ is known or can be estimated directly from the training set.
3. $p(\omega_i|\underline{x})$ is generally not known.

Here $p(\omega_i)$ is the a priori probability of class ω_i, and $p(\underline{x}|\omega_i)$ is the likelihood function of class ω_i, or the state conditional probability density function of $\underline{x}$. More explicitly, it is the probability density function for $\underline{x}$ given that the state of nature is ω_i and $p(\omega_i|\underline{x})$ is the probability that $\underline{x}$ comes from ω_i. This is actually the a posteriori probability.

A loss function L_{ij} may be defined as the loss, cost, or penalty due to deciding that $\underline{x} \in \omega_j$ when, in fact, $\underline{x} \in \omega_i$. Thus we seek to minimize the average loss. Similarly, the conditional average loss or conditional average risk $r_k(\underline{x})$ may be defined as

$$r_k(\underline{x}) = \sum_{i=1}^{M} L_{ik}p(\omega_i|\underline{x}) \qquad (4.1)$$

that is, the average or expected loss of misclassifying $\underline{x}$ as in ω_k; but in fact, it should be in some other classes ω_i, $i = 1, 2, \ldots, M$ and $i \neq k$.

The job of the classifier is then to find an optimal decision that will minimize the average risk or cost. The decision rule will then consist of the following steps:

Step 1: Compute the expected losses, $r_i(\underline{x})$, of deciding that $\underline{x} \in \omega_i$ ∀ i, i = 1, 2, ..., M.
Step 2: Decide that $\underline{x} \in \omega_k$ if $r_k(\underline{x}) \leq r_i(\underline{x})$ ∀ i, i ≠ k.

The corresponding discriminant function is then

$$d_k(\underline{x}) = -r_k(\underline{x}) \qquad (4.2)$$

The negative sign in front of $r_k(\underline{x})$ is chosen so as to make $d_k(\underline{x})$ represent the most likely class. The smaller $r_k(\underline{x})$, the more likely it is that $\underline{x} \in \omega_k$.

A loss matrix can then be set up as

$$
\underline{L} =
\begin{bmatrix}
0 & & & & 1 \\
 & 0 & & & \\
 & & \cdot & & \\
 & & & \cdot & \\
1 & & & \cdot & \\
 & & & & 0
\end{bmatrix}
\tag{4.3}
$$

where $L_{ii} = 0$, $i = 1, \ldots, M$, since no misclassifications occur in such cases; while for $L_{ik} = 1$, there is a penalty in misclassifying $\underline{x} \in \omega_k$, but actually $\underline{x} \in \omega_i$, $i = 1, \ldots, M$, $i \neq k$. This is a symmetric loss function, since

$$
L_{ik} = 1 - \delta(k - i)
\tag{4.4}
$$

where $\delta(k - i)$ is the Kronecker delta function and

$$
\delta(k - i) =
\begin{cases}
1 & \text{if } k = i \\
0 & \text{otherwise}
\end{cases}
\tag{4.5}
$$

If the value of L_{ik} is such that

$$
L_{ik} =
\begin{cases}
-h_i & i = k \\
0 & i \neq k
\end{cases}
\tag{4.6}
$$

The loss matrix becomes a negative loss function matrix:

$$
\underline{L}_{neg} =
\begin{bmatrix}
-h_1 & & & & & \\
 & -h_2 & & & 0 & \\
 & & -h_3 & & & \\
 & & & \cdot & & \\
 & 0 & & & \cdot & \\
 & & & & & \cdot \\
 & & & & & -h_M
\end{bmatrix}
\tag{4.7}
$$

The significance of this negative loss function matrix is that a negative loss (i.e., a positive gain) is assigned to a correct decision and no loss to an erroneous decision. In other words, the loss assigned to a decision is greater for an erroneous decision than for a correct one.

Note that the h_i in the matrix do not have to be equal. They may be different to indicate the relative importance of guessing correctly on one class rather than the other. Similarly, the L_{ik} and L_{ki} in the loss matrix do not have to be equal.

For a two-class problem, $L_{ik} = L_{21}$, where $i = 2$, $k = 1$. This means that $\underline{x}$ should be in ω_2 but is misclassified as being in ω_1. $L_{ik} = L_{12}$ when $i = 1$, $k = 2$, meaning that $\underline{x}$ should be in ω_1 but is misclassified as being in ω_2. $L_{ik} = 0$ when $i = k$. Thus we have

$$\underline{L} = \begin{bmatrix} 0 & L_{21} \\ L_{12} & 0 \end{bmatrix} \tag{4.8}$$

Suppose that ω_1 is the class of friendly aircraft and ω_2 is the class of enemy aircraft; then undoubtedly,

$$L_{21} > L_{12}$$

since L_{12} is just a false alarm, but L_{21} would mean disaster.

However, in another example, such as a fire sprinkling system in a laboratory with expensive equipment, a false alarm should have a large L, because when the sprinkler goes off and there is no fire, a lot of equipment could be ruined. Thus we may end up with $L_{12} = L_{21}$, which is then a symmetric loss function.

4.2.2 Bayes' Discriminant Function

By Bayes' rule, we can write

$$p(\omega_i|\underline{x}) = \frac{p(\underline{x}|\omega_i)p(\omega_i)}{p(\underline{x})} \tag{4.9}$$

where $p(\underline{x}) = \Sigma_i p(\underline{x}|\omega_i)p(\omega_i)$, $i = 1, 2, \ldots, M$, is the probability that $\underline{x}$ occurs without regard to the category in which it belongs. $p(\omega_i)$ is the a priori probability of class ω_i, and $p(\underline{x}|\omega_i)$ is the likelihood function of class ω_i with respect to $\underline{x}$; it is the probability density function for $\underline{x}$ given that the state of nature is ω_i (i.e., it is a pattern belonging to class ω_i).

Substituting Eq. (4.9) into (4.1) for $r_k(\underline{x})$, we have

$$r_k(\underline{x}) = \frac{1}{p(\underline{x})} \sum_{i=1}^{M} L_{ik}p(\underline{x}|\omega_i)p(\omega_i) \tag{4.10}$$

Since $p(\underline{x})$ in Eq. (4.10) is common to all $r_j(\underline{x})$, $j = 1, \ldots, M$, we can eliminate it from the conditional average risk equation and seek only the following minimum:

$$\min_{k} \, r_k(\underline{x}) = \min_{k} \sum_{i=1}^{M} L_{ik} p(\underline{x}|\omega_i) p(\omega_i) \tag{4.11}$$

to obtain the best one among all the possible decisions, or alternatively, we can just say that

$$d_k(\underline{x}) = -r_k(\underline{x}) \tag{4.12}$$

which is the Bayes discriminant function. The classifier basing on this minimization is called Bayes' classifier, which gives the optimum performance from the statistical point of view.

4.2.3 Maximum Likelihood Decision

As defined in Sec. 4.2.2, $p(\underline{x}|\omega_i)$ is called the likelihood function of ω_i. The expression for average or expected loss of deciding $\underline{x} \in \omega_k$ is

$$r_k(\underline{x}) = \sum_{i=1}^{M} L_{ik} p(\underline{x}|\omega_i) p(\omega_i) \tag{4.13}$$

which can then be used for minimization to get the maximum likelihood for $\underline{x} \in \omega_k$. For a two-class problem, the average or expected loss of deciding $\underline{x} \in \omega_1$ will be

$$r_1(\underline{x}) = L_{11} p(\underline{x}|\omega_1) p(\omega_1) + L_{21} p(\underline{x}|\omega_2) p(\omega_2) \tag{4.14}$$

Similarly, the loss of deciding $\underline{x} \in \omega_2$ will be

$$r_2(\underline{x}) = L_{12} p(\underline{x}|\omega_1) p(\omega_1) + L_{22} p(\underline{x}|\omega_2) p(\omega_2) \tag{4.15}$$

In matrix form,

$$\underline{r} = \underline{L}\underline{p} \tag{4.16}$$

or

$$\begin{vmatrix} r_1 \\ r_2 \end{vmatrix} = \begin{vmatrix} L_{11} & L_{21} \\ L_{12} & L_{22} \end{vmatrix} \begin{vmatrix} p(\underline{x}|\omega_1) p(\omega_1) \\ p(\underline{x}|\omega_2) p(\omega_2) \end{vmatrix} \tag{4.17}$$

The decision that $\underline{x} \in \omega_1$ will be made if

$$L_{11}p(\underline{x}|\omega_1)p(\omega_1) + L_{21}p(\underline{x}|\omega_2)p(\omega_2)$$
$$< L_{12}p(\underline{x}|\omega_1)p(\omega_1) + L_{22}p(\underline{x}|\omega_2)p(\omega_2)$$

(4.18)

or

$$(L_{21} - L_{22})p(\underline{x}|\omega_2)p(\omega_2) < (L_{12} - L_{11})p(\underline{x}|\omega_1)p(\omega_1) \qquad (4.19)$$

The inequality above can be put in another form; that is, we assign $\underline{x} \in \omega_1$ if

$$\frac{p(\underline{x}|\omega_1)}{p(\underline{x}|\omega_2)} > \frac{(L_{21} - L_{22})p(\omega_2)}{(L_{12} - L_{11})p(\omega_1)} \qquad (4.20)$$

Using the notation $\ell_{12}(\underline{x})$ for $p(\underline{x}|\omega_1)/p(\underline{x}|\omega_2)$ as the likelihood ratio and θ_{12} for $(L_{21} - L_{22})p(\omega_2)/(L_{12} - L_{11})p(\omega_1)$ as the threshold value, the criterion for the decision becomes

$$\underline{x} \in \omega_1 \quad \text{if} \quad \ell_{12}(\underline{x}) > \theta_{12} \qquad (4.21)$$

The derivation above can easily be generalized to a multiclass problem (i.e., when $M > 2$). The generalized likelihood ratio and the generalized threshold value will become, respectively,

$$\ell_{ki} = \frac{p(\underline{x}|\omega_k)}{p(\underline{x}|\omega_i)} \qquad (4.22)$$

and

$$\theta_{ki} = \frac{(L_{ik} - L_{ii})p(\omega_i)}{(L_{ki} - L_{kk})p(\omega_k)} \qquad (4.23)$$

Then the criterion for the decision can be stated:

$$\text{Assign } \underline{x} \in \omega_k \quad \text{if} \quad \ell_{ki} > \theta_{ki} \quad \forall \; i \qquad (4.24)$$

This is what we call the *maximum likelihood rule*. Basing on these mathematical relations, it would not be difficult to implement it as a classifier.

Now let us consider the case that L is a symmetric loss function. The problem becomes to assign $\underline{x} \in \omega_k$ if $\ell_{ki} \geqslant \theta_{ki}$ $\forall$ i, i = 1, ..., M. The maximum likelihood rule for this symmetric loss function becomes

$$\frac{p(\underline{x}|\omega_k)}{p(\underline{x}|\omega_i)} \geqslant \frac{p(\omega_i)}{p(\omega_k)} \tag{4.25}$$

since $L_{ik} = 1$ and $L_{ii} = 0$ $\forall$ i, k and $i \neq k$; i, k = 1, ..., M. If $p(\omega_i) = p(\omega_k)$ $\forall$ i, k, the maximum likelihood rule becomes:

$$\text{Assign } \underline{x} \in \omega_k \quad \text{if } \ell_{ki} \geqslant 1 \tag{4.26}$$

Note that a different loss function yields a different maximum likelihood rule.

Now let us go back to the more general case that $p(\omega_i) \neq p(\omega_k)$, and let us formulate a discriminant function for the case of the symmetric loss function. Since we have

$$\ell_{ki} = \frac{p(\underline{x}|\omega_k)}{p(\underline{x}|\omega_i)} \geqslant \frac{p(\omega_i)}{p(\omega_k)} \tag{4.27}$$

then

$$p(\underline{x}|\omega_k)p(\omega_k) \geqslant p(\underline{x}|\omega_i)p(\omega_i) \tag{4.28}$$

In other words, we can assign $\underline{x} \in \omega_k$ if

$$p(\underline{x}|\omega_k)p(\omega_k) \geqslant p(\underline{x}|\omega_i)p(\omega_i) \quad \forall \quad i \tag{4.29}$$

Thus the discriminant function is now

$$d_k(\underline{x}) = p(\underline{x}|\omega_k)p(\omega_k) \tag{4.30}$$

An alternative form of this discriminent function is

$$d_k'(\underline{x}) = \log p(\underline{x}|\omega_k) + \log p(\omega_k) \tag{4.31}$$

Extending this to a more general case, the average loss of deciding that $\underline{x} \in \omega_k$ is

$$r_k(\underline{x}) = \sum_{i=1}^{M} L_{ik}p(\underline{x}|\omega_i)p(\omega_i) \tag{4.32}$$

or

$$\underline{r} = \underline{L}^T \underline{p} \qquad (4.33)$$

The maximum likelihood rule is then

$$\underline{x} \in \omega_i \quad \text{if } r_i(\underline{x}) < r_j(\underline{x}) \qquad (4.34)$$

or

$$\sum_{k=1}^{M} L_{ki} p(\underline{x}|\omega_k) p(\omega_k) < \sum_{q=1}^{M} L_{qj} p(\underline{x}|\omega_q) p(\omega_q)$$

$$\forall \quad j, \ j \neq i; \ j = 1, \ldots, M \qquad (4.35)$$

The summation of the terms on the left-hand side of (4.35) represent the average loss of deciding $\underline{x} \in \omega_i$, while that on the right-hand side represents the loss of deciding that $\underline{x} \in \omega_j$, $j = 1, \ldots, M$ and $j \neq i$.

4.2.4 Binary Example

Let us take, for illustration, a binary example, in which each pattern $\underline{x}$ has independent binary components as follows:

$$\underline{x} = [x_1, x_2, \ldots, x_n]^T \quad x_i = 1 \text{ or } 0, \ i = 1, 2, \ldots, n \qquad (4.36)$$

For a two-class problem ($M = 2$), the discriminant function $d(\underline{x})$ is

$$d(\underline{x}) = d_1(\underline{x}) - d_2(\underline{x}) \qquad (4.37)$$

where $d_1(\underline{x}) = \log [p(\underline{x}|\omega_1)p(\omega_1)]$ and $d_2(\underline{x}) = \log [p(\underline{x}|\omega_2)p(\omega_2)]$. Then

$$d(\underline{x}) = \log \frac{p(\underline{x}|\omega_1)}{p(\underline{x}|\omega_2)} + \log \frac{p(\omega_1)}{p(\omega_2)} \qquad (4.38)$$

For a two-class problem, $p(\omega_1) + p(\omega_2) = 1$; hence

$$d(\underline{x}) = \log \frac{p(\underline{x}|\omega_1)}{p(\underline{x}|\omega_2)} + \log \frac{p(\omega_1)}{1 - p(\omega_1)} \qquad (4.39)$$

Since the components x_i are independent,

$$p(\underline{x}|\omega_j) = p(x_1|\omega_j)p(x_2|\omega_j) \cdots p(x_n|\omega_j) = \prod_{i=1}^{n} p(x_i|\omega_j) \qquad (4.40)$$

and

$$d(\underline{x}) = \sum_{i=1}^{n} \log \frac{p(x_i|\omega_1)}{p(x_i|\omega_2)} + \log \frac{p(\omega_1)}{1 - p(\omega_1)} \qquad (4.41)$$

Since the pattern elements of $\underline{x}$ are binary for the problem we discussed here, for simplicity we can let

$$p(x_i = 1|\omega_1) = p_i \qquad (4.42)$$

Then

$$p(x_i = 0|\omega_1) = 1 - p_i \qquad (4.43)$$

Similarly, let

$$p(x_i = 1|\omega_2) = q_i \qquad (4.44)$$

Then

$$p(x_i = 0|\omega_2) = 1 - q_i \qquad (4.45)$$

We can then claim that

$$\log \frac{p(x_i|\omega_1)}{p(x_i|\omega_2)} = x_i \log \frac{p_i}{q_i} + (1 - x_i) \log \frac{1 - p_i}{1 - q_i} = \gamma \qquad (4.46)$$

The validity of (4.46) can easily be checked by setting either $x_i = 1$ or 0 in this expression. Rewriting expression (4.46) gives

$$\gamma = \log \frac{p(x_i|\omega_1)}{p(x_i|\omega_2)} = x_i \log \frac{p_i(1 - q_i)}{q_i(1 - p_i)} + \log \frac{1 - p_i}{1 - q_i} \qquad (4.47)$$

Substituting back in Eq. (4.41) gives

$$d(\underline{x}) = \sum_{i=1}^{n} x_i \log \frac{p_i(1 - q_i)}{q_i(1 - p_i)} + \sum_{i=1}^{n} \log \frac{1 - p_i}{1 - q_i} + \log \frac{p(\omega_1)}{1 - p(\omega_1)}$$

$$(4.48)$$

where $\log[p_i(1 - q_i)/q_i(1 - p_i)]$ can be represented by w_i and

$$\sum_{i=1}^{n} \log \frac{1 - p_i}{1 - q_i} + \log \frac{p(\omega_1)}{1 - p(\omega_1)}$$

can be represented by w_{n+1}. Then we have

$$d(\underline{x}) = \sum_{i=1}^{n} w_i x_i + w_{n+1} \qquad (4.49)$$

from which we can see that the optimum discriminant function is linear in $\underline{x_i}$.

4.2.5 Probability of Error

The probability of error that would be introduced in the scheme discussed in Sec. 4.2.4 is a problem of much concern. Let us again take the two-class problem for illustration. The classifier will divide the space into two regions, R_1 and R_2. The decision that $\underline{x} \in \omega_1$ will be made when the pattern $\underline{x}$ falls into the region R_1; and $\underline{x} \in \omega_2$, when $\underline{x}$ falls into R_2. Under such circumstances, there will be two possible types of errors:

1. $\underline{x}$ falls in region R_1, but actually $\underline{x} \in \omega_2$. This gives the probability of error E_1, which may be denoted by Prob($\underline{x} \in R_1$, ω_2).
2. $\underline{x}$ falls in region R_2, but actually $\underline{x} \in \omega_1$. This gives the probability of error E_2, or Prob($\underline{x} \in R_2$, ω_1). Thus the total probability of error is

$$P_{error} = \text{Prob}(\underline{x} \in R_1 | \omega_2)p(\omega_2) + \text{Prob}(\underline{x} \in R_2 | \omega_1)p(\omega_1)$$

$$= \int_{R_1} p(\underline{x} | \omega_2)p(\omega_2) \, d\underline{x} + \int_{R_2} p(\underline{x} | \omega_1)p(\omega_1) \, d\underline{x}$$

$$(4.50)$$

This is the performance criterion that we try to minimize to give a good classification. The two integrands in expression (4.50) are plotted in Fig. 4.1.

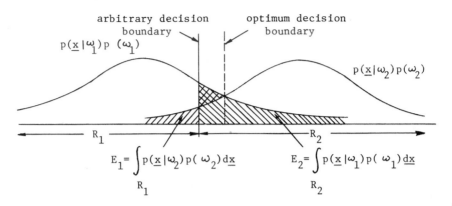

FIG. 4.1 Probability of error in a two-class problem.

Areas under the curves shown by the hatched lines represent E_1 and E_2, where $E_1 = \int_{R_1} p(\underline{x}|\omega_2)p(\omega_2) \, d\underline{x}$ and $E_2 = \int_{R_2} p(\underline{x}|\omega_1)$ $p(\omega_1) \, d\underline{x}$. It is not difficult to see that with an arbitrary decision boundary, E_2 represents both the right slash-hatched and the cross-hatched areas. If the decision boundary is moved to the right to the optimum position, which is the vertical line passing through the intersection of the two probability curves, the double-hatched area is eliminated from the total area and a minimum error would occur. This optimum decision boundary occurs when $\underline{x}$ satisfies the following equation:

$$d_1(\underline{x}) = d_2(\underline{x}) \tag{4.51}$$

or

$$p(\underline{x}|\omega_1)p(\omega_1) = p(\underline{x}|\omega_2)p(\omega_2) \tag{4.52}$$

and hence the maximum likelihood rule (or Bayes' decision rule with symmetric loss function) is the optimum classifier from a minimum probability of error viewpoint.

To give an analytical expression for the probability of error, let us assume multivariate normal density functions for the pattern vectors with $\underline{C}_1 = \underline{C}_2 = \underline{C}$; thus

$$p(\underline{x}|\omega_1) = \frac{1}{(2\pi)^{n/2}|\underline{C}|^{1/2}} \exp\left[-\frac{1}{2}(\underline{x} - \underline{m}_1)^T \underline{C}^{-1}(\underline{x} - \underline{m}_1) \right]$$

$$\tag{4.53}$$

and

$$p(\underline{x}|\omega_2) = \frac{1}{(2\pi)^{n/2}|\underline{C}|^{1/2}} \exp\left[-\frac{1}{2}(\underline{x} - \underline{m}_2)^T \underline{C}^{-1}(\underline{x} - \underline{m}_2)\right]$$

(4.54)

Then, according to Eqs. (4.20) and (4.21),

$$\underline{x} \in \omega_1 \in \text{ if } \ell_{12} > \theta_{12}$$

(4.55)

or

$$\frac{p(\underline{x}|\omega_1)}{p(\underline{x}|\omega_2)} > \frac{(L_{21} - L_{22})p(\omega_2)}{(L_{12} - L_{11})p(\omega_1)}$$

(4.56)

For the case where the loss functions are symmetric, we have

$$\frac{p(\underline{x}|\omega_1)}{p(\underline{x}|\omega_2)} > \frac{p(\omega_2)}{p(\omega_1)} = \theta'_{12}$$

(4.57)

Similarly, $\underline{x} \in \omega_2$ if $\ell_{21} > \theta_{21}$, that is,

$$\underline{x} \in \omega_2 \quad \text{if } \frac{p(\underline{x}|\omega_2)}{p(\underline{x}|\omega_1)} > \frac{p(\omega_1)}{p(\omega_2)} = \theta'_{21}$$

(4.58)

Substituting the normal density functions for $p(\underline{x}|\omega_1)$ and $p(\underline{x}|\omega_2)$, respectively, we obtain

$$\frac{p(\underline{x}|\omega_1)}{p(\underline{x}|\omega_2)} = \frac{\exp[-(1/2)(\underline{x} - \underline{m}_1)^T \underline{C}^{-1}(\underline{x} - \underline{m}_1)]}{\exp[-(1/2)(\underline{x} - \underline{m}_2)^T \underline{C}^{-1}(\underline{x} - \underline{m}_2)]}$$

(4.59)

Taking the logarithm of the ratio $p(\underline{x}|\omega_1)/p(\underline{x}|\omega_2)$ and denoting it by P_{12}, then

$$P_{12} = -\frac{1}{2}(\underline{x} - \underline{m}_1)^T \underline{C}^{-1}(\underline{x} - \underline{m}_1) + \frac{1}{2}(\underline{x} - \underline{m}_2)^T \underline{C}^{-1}(\underline{x} - \underline{m}_2)$$

$$= \underline{x}^T \underline{C}^{-1}(\underline{m}_1 - \underline{m}_2) - \frac{1}{2}(\underline{m}_1 + \underline{m}_2)^T \underline{C}^{-1}(\underline{m}_1 - \underline{m}_2)$$

(4.60)

Then

$$\text{Prob}(\underline{x} \in R_1, \, \omega_2) = p[p_{12} > \log \theta'_{12} | \omega_2] \tag{4.61}$$

and

$$\text{Prob}(\underline{x} \in R_2, \, \omega_1) = p[p_{12} < \log \theta'_{12} | \omega_1] \tag{4.62}$$

The expected value of p_{12} for class 1 can then be found as

$$E_1[p_{12}] = \underline{m}_1^T \underline{C}^{-1}(\underline{m}_1 - \underline{m}_2) - \frac{1}{2}(\underline{m}_1 + \underline{m}_2)^T \underline{C}^{-1}(\underline{m}_1 - \underline{m}_2)$$

$$= \frac{1}{2}[(\underline{m}_1 - \underline{m}_2)^T \underline{C}^{-1}(\underline{m}_1 - \underline{m}_2)] \tag{4.63}$$

The variance of p_{12} for class 1 is defined by

$$\text{var}_1[p_{12}] = E_1[(p_{12} - \overline{p_{12}})^2] \tag{4.64}$$

and equals

$$\text{var}_1[p_{12}] = E_1[\underline{x}^T \underline{C}^{-1}(\underline{m}_1 - \underline{m}_2) - \frac{1}{2}(\underline{m}_1 + \underline{m}_2)^T \underline{C}^{-1}(\underline{m}_1 - \underline{m}_2)$$

$$- \underline{m}_1^T \underline{C}^{-1}(\underline{m}_1 - \underline{m}_2) + \frac{1}{2}(\underline{m}_1 + \underline{m}_2)^T \underline{C}^{-1}(\underline{m}_1 - \underline{m}_2)]^2$$

$$= E_1[(\underline{x} - \underline{m}_1)^T \underline{C}^{-1}(\underline{m}_1 - \underline{m}_2)]^2$$

$$= E_1[(\underline{m}_1 - \underline{m}_2)^T \underline{C}^{-1}(\underline{x} - \underline{m}_1)(\underline{x} - \underline{m}_1)^T \underline{C}^{-1}(\underline{m}_1 - \underline{m}_2)]$$

$$= (\underline{m}_1 - \underline{m}_2)^T \underline{C}^{-1} E_1[(\underline{x} - \underline{m}_1)(\underline{x} - \underline{m}_1)^T \underline{C}^{-1}(\underline{m}_1 - \underline{m}_2)]$$

$$= (\underline{m}_1 - \underline{m}_2)^T \underline{C}^{-1} \underline{CC}^{-1}(\underline{m}_1 - \underline{m}_2) \tag{4.65}$$

since $E_1[(\underline{x} - \underline{m}_1)(\underline{x} - \underline{m}_1)^T] = \underline{C}$ by definition. Therefore.

$$\text{var}_1[p_{12}] = (\underline{m}_1 - \underline{m}_2)^T \underline{C}^{-1}(\underline{m}_1 - \underline{m}_2)$$

$$= r_{12} \tag{4.66}$$

Substituting back in Eq. (4.63), we obtain

$$E_1[p_{12}] = \frac{1}{2} r_{12} \tag{4.67}$$

where r_{12} equals the Mahalanobis distance between $p(\underline{x}|\omega_1)$ and $p(\underline{x}|\omega_2)$. Thus, for $\underline{x} \in \omega_1$, the ratio $p(\underline{x}|\omega_1)/p(\underline{x}|\omega_2)$ is distributed with a mean equal to $(1/2)r_{12}$ and a variance equal to r_{12}; while for $\underline{x} \in \omega_2$, that ratio will be distributed with a mean equal to $-(1/2)r_{12}$ and a variance equal to r_{12}. Therefore, the probability of misclassifying a pattern when $\underline{x} \in \omega_2$ is

$$p(p_{12} > \log \theta'_{12}|\omega_2) = \int_{\log\theta'_{12}}^{\infty} \frac{1}{\sqrt{2\pi r_{12}}} \exp\left\{-\frac{[p_{12} + (1/2)r_{12}]^2}{2r_{12}}\right\} dp_{12}$$

$$(4.68)$$

and the probability of misclassifying a pattern when it comes from ω_1 is

$$p(p_{12} < \log \theta'_{12}|\omega_1) = \int_{-\infty}^{\log\theta'_{12}} \frac{1}{\sqrt{2\pi r_{12}}}$$

$$\exp\left\{-\frac{[p_{12} - (1/2)r_{12}]^2}{2r_{12}}\right\} dp_{12} \qquad (4.69)$$

The total probability of error, P_{error}, is then

$$P_{error} = E_1 + E_2$$

$$= p(p_{12} > \log \theta'_{12}|\omega_2)p(\omega_2) + p(p_{12} < \log \theta'_{12}|\omega_1)p(\omega_1)$$

$$(4.70)$$

This analysis can easily be extended to a multiclass case. In the multiclass cases, there are more ways to be wrong than to be right. So it is simpler to compute the probability of being correct.

Let us denote the probability of being correct as

$$P_{correct} = \sum_{i=1}^{M} Prob(\underline{x} \in R_i, \omega_i) = \sum_{i=1}^{M} \int_{R_i} p(\underline{x}|\omega_i)p(\omega_i)\, d\underline{x}$$

$$(4.71)$$

where $\text{Prob}(\underline{x} \in R_i, \omega_j)$ denotes the probability that $\underline{x}$ falls in R_i, while the true state of nature is also that $\underline{x} \in \omega_j$. Summation of $\text{Prob}(\underline{x} \in R_i, \omega_j)$, $i = 1, 2, \ldots, M$, gives the total classification probability of being correct. The Bayes classifier with symmetric loss function maximizes P_{correct} by choosing the regions R_i so that the integrands are maximum. Analysis of the multivariate normal density function for the pattern vectors can be worked out similarly without too much difficulty.

4.3 OPTIMAL DISCRIMINANT FUNCTIONS FOR NORMALLY DISTRIBUTED PATTERNS

4.3.1 Normal Distribution

The multivariate normal density function for M pattern classes can be represented by

$$p(\underline{x}|\omega_k) = \frac{1}{(2\pi)^{n/2}|\underline{C}_k|^{1/2}} \exp\left[-\frac{1}{2}(\underline{x} - \underline{m}_k)^T \underline{C}_k^{-1}(\underline{x} - \underline{m}_k)\right]$$

$$= N(\underline{m}_k, \underline{C}_k) \quad k = 1, 2, \ldots, M; \ n = \text{dimensionality of the pattern vector}$$

(4.72)

where N is the normal density function, $\underline{m}_k$ is the mean vector, and $\underline{C}_k$ is the covariance matrix for class k, defined respectively by their expected values over the patterns belonging to the class k. Thus

$$\underline{m}_k = E_k[\underline{x}]$$

(4.73)

and

$$\underline{C}_k = E_k[(\underline{x} - \underline{m}_k)(\underline{x} - \underline{m}_k)^T]$$

(4.74)

Pattern samples drawn from a normal population in the pattern space form a single cluster, the center of which is determined by the mean vector obtained from the samples and the shape of the cluster is determined by the covariance matrix. Figure 4.2 shows three different clusters with different shapes. For the cluster in part (a), $\underline{m} = 0$ and $\underline{C} = I$ (an identity matrix). Because of its symmetry, $C_{ji} = C_{ij} = 0$. $C_{ii} = 1$. For the cluster in part (b),

$$\underline{m} = \begin{bmatrix} 1 \\ 0 \end{bmatrix} \quad \text{and} \quad \underline{C} = \begin{bmatrix} C_{11} & 0 \\ 0 & C_{22} \end{bmatrix}$$

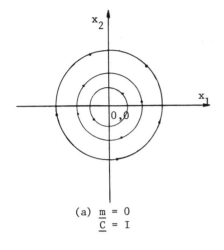

(a) $\underline{m} = 0$
$\underline{C} = I$

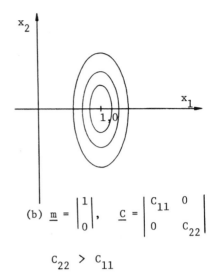

(b) $\underline{m} = \begin{vmatrix} 1 \\ 0 \end{vmatrix}$, $\underline{C} = \begin{vmatrix} C_{11} & 0 \\ 0 & C_{22} \end{vmatrix}$

$C_{22} > C_{11}$

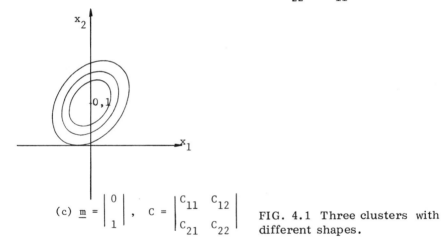

(c) $\underline{m} = \begin{vmatrix} 0 \\ 1 \end{vmatrix}$, $C = \begin{vmatrix} C_{11} & C_{12} \\ C_{21} & C_{22} \end{vmatrix}$

FIG. 4.1 Three clusters with different shapes.

$C_{22} > C_{11}$; while for the cluster in (c) (still in the same figure),

$$\underline{m} = \begin{bmatrix} 0 \\ 1 \end{bmatrix} \quad \text{and} \quad \underline{C} = \begin{bmatrix} C_{11} & C_{12} \\ C_{21} & C_{22} \end{bmatrix}$$

The principal axes of the hyperellipsoids (contours of equal probability density) are given by the eigenvectors of $\underline{C}$ with the eigenvalues determining the relative lengths of these axes.

A useful measure for similarity, known as *Mahalanobis distance* (r^2) from pattern $\underline{x}$ to mean $\underline{m}$, can be defined as

$$r^2 = (\underline{x} - \underline{m})^T C^{-1}(\underline{x} - \underline{m}) \tag{4.75}$$

The Mahalanobis distance between two classes can similarly be expressed as

$$r_{ij} = (\underline{m}_i - \underline{m}_j)^T C^{-1}(\underline{m}_i - \underline{m}_j) \tag{4.76}$$

Recall that for $n = 1$, approximately 95% of the samples $\underline{x}$ fall in the region $|\underline{x} - \underline{m}| < 2\sigma$, where σ is the standard deviation and is equal to $\underline{C}^{1/2}$.

4.3.2 Optimal Discriminant Functions

From Eq. (4.31), the discriminant function for $\underline{x} \in \omega_k$ can be put in the following form:

$$d'_k(\underline{x}) = \log p(\underline{x}|\omega_k) + \log p(\omega_k) \tag{4.77}$$

When this discriminant function is applied to the multivariate normal desnity for an M-pattern class with

$$p(\underline{x}|\omega_k) = \frac{1}{(2\pi)^{n/2}|\underline{C}_k|^{1/2}} \exp\left[-\frac{1}{2}(\underline{x} - \underline{m}_k)^T \underline{C}_k^{-1}(\underline{x} - \underline{m}_k)\right]$$

$$k = 1, 2, \ldots, M \tag{4.78}$$

the discriminant function $d'_k(\underline{x})$ becomes

$$d''_k(\underline{x}) = -\frac{n}{2}\log (2\pi) - \frac{1}{2}\log |\underline{C}_k| - \frac{1}{2}(\underline{x} - \underline{m}_k)^T \underline{C}_k^{-1}(\underline{x} - \underline{m}_k)$$

$$+ \log p(\omega_k) \tag{4.79}$$

It is clear that if the first term on the right-hand side is the same for all k, it can be eliminated. Then the discriminant function reduces to

$$d'''_k(\underline{x}) = -\frac{1}{2}(\underline{x} - \underline{m}_k)^T \underline{C}_k^{-1}(\underline{x} - \underline{m}_k) + [\log p(\omega_k) - \frac{1}{2}\log|\underline{C}_k|]$$

$$\tag{4.80}$$

This is a quadratic discriminant function, and can be put in more compact form as

$$d_k^{(iv)}(\underline{x}) = -\frac{1}{2}r^2 + f(k) \quad \text{for } \underline{x} \in \omega_k \quad (4.81)$$

where $r^2 = (\underline{x} - \underline{m}_k)^T C_k^{-1}(\underline{x} - \underline{m}_k)$ is the Mahalanobis distance defined by Eq. (4.75) and $f(k) = \log p(\omega_k) - (1/2) \log |C_k|$. Let us discuss this discriminant function in more detail for two different cases.

Case 1: When the Covariance Matrices Are Equal for Different Classes (i.e., $C_i = C_j = C_k = C$) The physical significance of this is that the separate classes (or clusters in our special terminology) are of equal size and of similar shape, but the clusters are centered about different means. Expanding the general equation for $d_k(\underline{x})$ [Eq. (4.80)], we get

$$d_k(\underline{x}) = -\frac{1}{2}\underline{x}^T C^{-1}\underline{x} - \frac{1}{2}\underline{m}_k^T C^{-1}\underline{m}_k + \underline{x}^T C^{-1}\underline{m}_k + \log p(\omega_k)$$
$$- \frac{1}{2}\log |C| \quad (4.82)$$

The first and last terms on the right-hand side of Eq. (4.82) are the same for all classes (i.e., for all k). Then this discriminant function can be put in an even more compact form as follows:

$$d_k(\underline{x}) = \underline{x}^T C^{-1}\underline{m}_k + [\log p(\omega_k) - \frac{1}{2}\underline{m}_k^T C^{-1}\underline{m}_k] \quad k = 1, 2, \ldots, M$$
$$(4.83)$$

Obviously, this is a linear discriminant function if we treat $C^{-1}\underline{m}_k$ as $\underline{w}_k$ and treat the two terms inside the brackets as an augmented term, $w_{k,n+1}$. For a two-class problem (M = 2),

$$d(\underline{x}) = d_1(\underline{x}) - d_2(\underline{x})$$

$$= \underline{x}^T C^{-1}(\underline{m}_1 - \underline{m}_2) + \log \frac{p(\omega_1)}{p(\omega_2)} - \frac{1}{2}(\underline{m}_1^T C^{-1}\underline{m}_1 - \underline{m}_2^T C^{-1}\underline{m}_2)$$
$$(4.84)$$

or

$$d(\underline{x}) = \underline{x}^T C^{-1}(\underline{m}_1 - \underline{m}_2) + \log \frac{p(\omega_1)}{p(\omega_2)} - \frac{1}{2}[(\underline{m}_1 - \underline{m}_2)^T C^{-1}(\underline{m}_1 + \underline{m}_2)]$$
$$(4.85)$$

Case 2: When the Covariance Matrix C_k Is of Diagonal Form $\sigma_k^2 I$, Where $\sigma_k^2 = |C_k|$ The physical significance of this is that the cluster has equal components along all the principal axes, and the distribution is of spherical shape. Then substitution of $\sigma_k^2 I$ for $\underline{C}_k$ in Eq. (4.80) gives

$$d_k(\underline{x}) = -\frac{1}{2} \frac{(\underline{x} - \underline{m}_k)^T (\underline{x} - \underline{m}_k)}{\sigma_k^2} + \left[\log p(\omega_k) - \frac{1}{2} \log \sigma_k^2 \right]$$

(4.86)

because $C_k^{-1} = (1/\sigma_k^2)I$. When the features are statistically independent, and when each feature has the same variance, σ^2, then $\sigma_k = \sigma_j = \sigma$ $\forall$ j, k, that is,

$$\underline{C}_k = \underline{C}_j = \sigma^2 I$$

(4.87)

and

$$d_k(\underline{x}) = -\frac{1}{2} \frac{\underline{x}^T \underline{x} - 2\underline{x}^T \underline{m}_k + \underline{m}_k^T \underline{m}_k}{\sigma^2} + \log p(\omega_k) - \frac{1}{2} \log \sigma^2$$

(4.88)

Again, $\underline{x}^T \underline{x}$ and $(1/2) \log \sigma^2$ are the same for all k. We can neglect these two terms in $d_k(\underline{x})$ and get a new expression:

$$d_k(\underline{x}) = \frac{1}{\sigma^2} \underline{x}^T \underline{m}_k + \left[\log p(\omega_k) - \frac{1}{2\sigma^2} \underline{m}_k^T \underline{m}_k \right]$$

(4.89)

which can then also be treated as a linear discriminant function.

If in addition to the assumption that $\underline{C}_k = \underline{C}_j = \sigma^2 I$, the assumption is made to let $p(\omega_k) = 1/K$ $\forall$ k, where K is a constant, then the term "$\log p(\omega_k)$" can also be dropped from the expression for $d_k(\underline{x})$. Then $d_k(\underline{x})$ will be further simplified as

$$d_k(\underline{x}) = \underline{x}^T \underline{m}_k - \frac{1}{2} |\underline{m}_k|^2$$

(4.90)

which is obviously a linear equation.

From the analyses we have done so far, the quadratic discriminant function as obtained for the multivariate normal density for M-pattern classes can be simplified into a form that can be implemented by a linear machine, thus making the problem much simpler.

Equation (4.86) can be simplified into another form, with which we are familiar. Since it is assumed that $C_k = C_j = C = \sigma^2 I$ and $p(\omega_k) = p(\omega_j) = \cdots = 1/K = $ constant, after dropping the unnecessary terms, Eq. (4.86) becomes

$$d_k(\underline{x}) = -\frac{1}{2} \frac{(\underline{x} - \underline{m}_k)^T (\underline{x} - \underline{m}_k)}{\sigma^2} \tag{4.91}$$

or simply

$$d_k(\underline{x}) = -(\underline{x} - \underline{m}_k)^T (\underline{x} - \underline{m}_k) = -|\underline{x} - \underline{m}_k|^2 \tag{4.92}$$

which is the same as the minimum distance classifier.

To conclude this section, we would like to add that the multivariate normal density function mentioned in Sec. 4.3.1 is only one of the probability density functions available to represent the distribution of random variables. If K_n, $|\underline{W}|^{1/2}$, and $f[(\underline{x} - \underline{m})^T \underline{W}(\underline{x} - \underline{m})]$ replace $(2\pi)^{n/2}$, $|\underline{C}|^{1/2}$, and $\exp[-(1/2)(\underline{x} - \underline{m})^T \underline{C}^{-1}(\underline{x} - \underline{m}]$, respectively, the multivariate normal density function

$$p(\underline{x}) = \frac{1}{(2\pi)^{n/2}|\underline{C}|^{1/2}} \exp\left[-\frac{1}{2}(\underline{x} - \underline{m})^T \underline{C}^{-1}(\underline{x} - \underline{m}) \right] \tag{4.93}$$

can be generalized as

$$p(\underline{x}) = K_n |\underline{W}|^{1/2} f[(\underline{x} - \underline{m})^T \underline{W}(\underline{x} - \underline{m})] \tag{4.94}$$

with K_n as the normalizing constant and $\underline{W}$ as the weight matrix. When different values and functions are given to K_n, $\underline{W}$, and f, different types of density function will be obtained. Examples of these are Pearson Type II and Type VII functions.

A very simple example is used to illustrate the computation of the mean, covariance, and the discriminant function by the statistical decision method. A practical example using computer computation with a large data set is given at the end of the chapter.

Example Given pattern points $(1,2)^T$, $(2,2)^T$, $(3,1)^T$, $(3,2)^T$, and $(2,3)^T$, are known to be in class ω_1. Another set of points, $(7,9)^T$, $(8,9)^T$, $(9,8)^T$, $(9,9)^T$, and $(8,10)^T$, are known to be in class ω_2. It is required to find a Bayes decision boundary to separate them.

Solution:

$$\underline{m}_1 = \frac{1}{N_1} \sum_{j=1}^{N_1} x_{1j} = \frac{1}{5} \begin{bmatrix} 11 \\ 10 \end{bmatrix}$$

$$\underline{m}_2 = \frac{1}{N_2} \sum_{j=1}^{N_2} x_{2j} = \frac{1}{5} \begin{bmatrix} 41 \\ 45 \end{bmatrix}$$

By definition,

$$\underline{C} = E[(\underline{x} - \underline{m})(\underline{x} - \underline{m})^T]$$

$$= E[\underline{x}\underline{x}^T] - \underline{m}\underline{m}^T$$

When it is put in discrete form,

$$\underline{C}_1 = \frac{1}{N_1} \sum_{j=1}^{N_1} x_{1j} x_{1j}^T - m_1 m_1^T$$

Therefore,

$$\underline{C}_1 = \frac{1}{5}\left[\begin{pmatrix} 1 \\ 2 \end{pmatrix} (1\ 2) + \begin{pmatrix} 2 \\ 2 \end{pmatrix} (2\ \ 2) + \begin{pmatrix} 3 \\ 1 \end{pmatrix} (3\ \ 1) + \begin{pmatrix} 3 \\ 2 \end{pmatrix} (3\ \ 2) \right.$$

$$\left. + \begin{pmatrix} 2 \\ 3 \end{pmatrix} (2\ \ 3) \right] - \frac{1}{25} \begin{pmatrix} 11 \\ 10 \end{pmatrix} (11\ \ 10)$$

$$= \frac{1}{25} \begin{pmatrix} 14 & -5 \\ -5 & 10 \end{pmatrix}$$

Similarly,

$$\underline{C}_2 = \frac{1}{N_2} \sum_{j=1}^{N_2} x_{2j} x_{2j}^T - \underline{m}_2 \underline{m}_2^T$$

$$= \frac{1}{25} \begin{pmatrix} 14 & -5 \\ -5 & 10 \end{pmatrix}$$

We have

$$\underline{C}_1 = \underline{C}_2 = \underline{C} = \frac{1}{25} \begin{pmatrix} 14 & -5 \\ -5 & 10 \end{pmatrix}$$

The determinant and adjoint of $\underline{C}$ can be computed as

$$|C| = \frac{1}{25} \begin{vmatrix} 14 & -5 \\ -5 & 10 \end{vmatrix} = \frac{23}{5} \qquad \text{Adj } \underline{C} = \begin{pmatrix} \dfrac{10}{25} & \dfrac{5}{25} \\[2mm] \dfrac{5}{25} & \dfrac{15}{25} \end{pmatrix}$$

The inverse of $\underline{C}$, $\underline{C}^{-1}\underline{m}_1$, and $\underline{m}_1^T \underline{C}^{-1}\underline{m}_1$ are then, respectively, as follows:

$$C^{-1} = \frac{1}{|\underline{C}|} \text{ adj } \underline{C} = \frac{1}{115} \begin{pmatrix} 10 & 5 \\ 5 & 14 \end{pmatrix}$$

$$\underline{C}^{-1}\underline{m}_1 = \frac{1}{115} \begin{pmatrix} 32 \\ 39 \end{pmatrix}$$

$$\underline{m}_1^T C^{-1}\underline{m}_1 = \frac{742}{5 \times 115}$$

The discriminant function for class 1 is

$$d_1(\underline{x}) = \underline{x}^T \underline{C}^{-1}\underline{m}_1 - \frac{1}{2}\underline{m}_1^T C^{-1}\underline{m}_1$$

$$= \frac{32}{115} x_1 + \frac{39}{115} x_2 - 0.65$$

Similarly, we obtain

$$\underline{C}^{-1}\underline{m}_2 = \frac{1}{115} \begin{pmatrix} 127 \\ 167 \end{pmatrix}$$

$$\underline{m}_2^T C^{-1}\underline{m}_2 = 22$$

The discriminant function for class 2 is

$$d_2(\underline{x}) = \underline{x}^T \underline{C}^{-1}\underline{m}_2 - \frac{1}{2}\underline{m}_2^T \underline{C}^{-1}\underline{m}_2$$

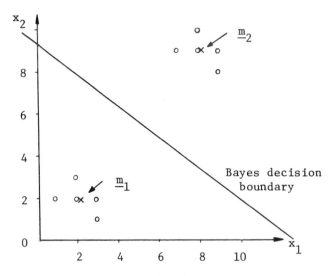

FIG. 4.3 Illustrative example.

$$= \frac{127}{115} x_1 + \frac{167}{115} x_2 - 11$$

The decision surface is then given by

$$d(\underline{x}) = d_1(\underline{x}) - d_2(\underline{x}) = 0$$

or

$$d(\underline{x}) = -0.826x_1 - 1.11x_2 + 10.35 = 0$$

which is shown in Fig. 4.3

4.4 TRAINING FOR STATISTICAL DISCRIMINANT FUNCTIONS

So far the formulation of the statistical classification problem and the optimum discriminant function for a normally distributed pattern have been discussed. The next problem that might interest us will be how to determine the unknown probability density function. One of the ways of doing this is by functional approximation. Assume that we wish to approximate $p(\underline{x}|\omega_i)$ by a set of functions

$$\hat{p}(\underline{x}|\omega_i) = \sum_{k=1}^{K} c_{ik}\phi_k(\underline{x}) \tag{4.95}$$

where the caret sign over $p(\cdot)$ represents the estimated value. The $\phi_k(\underline{x})$ are arbitrary functions and can be a set of some basic functions, which may be Hermite polynomials or others. The problem that we have now becomes to seek the coefficients c_{ik} so that the mean squared error

$$Q = \int_{\underline{x}} [p(\underline{x}|\omega_i) - \hat{p}(\underline{x}|\omega_i)]^2 \, d\underline{x} \tag{4.96}$$

over all $\underline{x}$ for class ω_i can be minimized. After substitution of Eq. (4.95) in Eq. (4.96), we have

$$Q = \int_{\underline{x}} [p(\underline{x}|\omega_i) - \sum_{k=1}^{K} c_{ik}\phi_k(\underline{x})]^2 \, d\underline{x} \tag{4.97}$$

A necessary condition for minimum Q is

$$\frac{\partial Q}{\partial c_{ik}} = 0 \quad k = 1, \ldots, K \tag{4.98}$$

or

$$\frac{\partial Q}{\partial c_{ik}} = 2 \int_{\underline{x}} \left[p(\underline{x}|\omega_i) - \sum_{k=1}^{K} c_{ik}\phi_k(\underline{x}) \right] \phi_k(\underline{x}) \, d\underline{x} = 0 \tag{4.99}$$

from which we get

$$\int_{\underline{x}} \phi_k(\underline{x}) \left[\sum_{k=1}^{K} c_{ik}\phi_k(\underline{x}) \right] \, d\underline{x} = \int_{\underline{x}} \phi_k(\underline{x})p(\underline{x}|\omega_i) \, d\underline{x} \tag{4.100}$$

Since, by definition, $\int_{\underline{x}} \phi_k(\underline{x})p(\underline{x}|\omega_i) \, d\underline{x}$ is the expected value $E_i[\phi_k(\underline{x})]$, then

$$\sum_{k=1}^{K} c_{ik} \int_{\underline{x}} \phi_k(\underline{x})\phi_k(\underline{x}) \, d\underline{x} = E_i[\phi_k(\underline{x})] \quad k = 1, \ldots, K \tag{4.101}$$

A set of K linear equations in c_{ik} ($k = 1, \ldots, K$) for a certain i can be obtained to solve for c_{ik}, but knowledge of $p(\underline{x}|\omega_i)$ is required.

Knowing that $E_i[\phi_k(\underline{x})] = \int_{\underline{x}} \phi_k(\underline{x})p(\underline{x}|\omega_i) \, d\underline{x}$ can be approximated by

$$E_i[\phi_k(\underline{x})] \simeq \frac{1}{N_i} \sum_{j=1}^{N_i} \phi_k(\underline{x}_j) \tag{4.102}$$

where N_i is the number of pattern samples in class i, then

$$\sum_{k=1}^{K} c_{ik} \int_{\underline{x}} \phi_k(\underline{x})\phi_k(\underline{x}) \, d\underline{x} = \frac{1}{N_i} \sum_{j=1}^{N_i} \phi_k(\underline{x}_j) \qquad k = 1, \ldots, K \tag{4.103}$$

This is a set of K linear equations and can be solved for the K $\phi_k(\underline{x})$'s. If, in particular, orthonormal functions are chosen for the $\phi_k(\underline{x})$'s, that is,

$$\int_{\underline{x}} \phi_i(\underline{x})\phi_j(\underline{x}) \, d\underline{x} = \begin{cases} 1 & i = j \\ 0 & i \neq j \end{cases} \tag{4.104}$$

then

$$c_{ik} = \frac{1}{N_i} \sum_{j=1}^{N_i} \phi_k(\underline{x}_j) \qquad \begin{array}{l} k = 1, \ldots, K \\ i = 1, \ldots, M \end{array} \tag{4.105}$$

Once the coefficients c_{ik} have been determined, the density function $\hat{p}(\underline{x}|\omega_i)$ is formed. Note that the $\underline{x}_j$ do not have to be stored but can be presented in sequential order. The c_{ik} can then be iteratively obtained from the following relation:

$$c_{ik}(N_i + 1) = \frac{1}{N_i + 1} [N_i c_{ik}(N_i) + \phi_k(\underline{x}_{N_i+1})] \tag{4.106}$$

where $c_{ik}(N_i)$ and $c_{ik}(N_i + 1)$ represent the coefficients obtained with N_i and $N_i + 1$ pattern samples, respectively. For more detailed discussion on this topic see Tou and Gonzalez, Pattern Recognition Principles.

4.5 APPLICATION TO A LARGE DATA-SET PROBLEM: A PRACTICAL EXAMPLE

Problems with large data sets are very common in our daily life. Many such problems can be found in agriculture, industry, and commerce as well as defense. An example consisting of three basic color bands (red, green, and blue), each with 254 × 606 digitizing picture elements from an aerial photograph of a water treatment plant area, is used for illustration. See Fig. 4.6 for the computer-processed image. Every pixel (picture element) in the image corresponds to a pattern point in the pattern space:

$$\underline{x}_{ij} = \begin{vmatrix} x_{ij1} \\ x_{ij2} \\ x_{ij3} \end{vmatrix}$$

The norm of the vector representing this pattern point is $\sqrt{\sum_{k=1}^{3} x_{ijk}^2}$, or

$$|x_{ij}| = (\underline{x}_{ij}^T \underline{x}_{ij})^{1/2} \qquad i = 1, 2, \ldots, 254; \ j = 1, 2, \ldots, 606$$

$\underline{x}_{ij}$ are computed and normalized to 256 gray levels for each of the three above channels.

The aerial image is mainly the photoreflection of the ground objects. From that image a fairly good idea can be obtained about what is in the image if a set of spectral responses, one for each spectral band, is chosen as the basis for analysis. (See Table 4.1)

A histogram of the data set gives us a rough idea of the gray-level distribution among the pixels. From the information given in the histogram, we can then set the appropriate gray-level limits to separate the pattern points into categories. A small set of known data (or

TABLE 4.1 Means and Standard Deviations for Red, Green, and Blue Spectral Data Sets of the Image

	Mean	Standard deviation
Red	33.80	17.84
Green	38.11	26.51
Blue	25.12	10.17

```
NUMBER OF OBSERVATIONS:    73602

UNNORMALIZED DATA USED

CHANNELS USED :  2  3  4

MEANS AND STANDARD DEVIATIONS FOR GIVEN CHANNELS
------------------------------------------------

CHANNEL      2       3       4
MEAN       33.60   38.11   25.12

ST. DEV.   17.16   26.51   10.17


VARIANCE-COVARIANCE MATRIX
--------------------------

          2       3       4
2      318.42
3      465.40  702.54
4      179.50  261.39  103.39


CORRELATION MATRIX FOR GIVEN CHANNELS
-------------------------------------

          2       3       4
2      1.00
3      0.98    1.00
4      0.99    0.97    1.00


           HISTOGRAM FOR CHANNEL  2    0.6 - 0.7    MICRONS.
           ------------------------------------------------

           EACH * REPRESENTS    47 OBSERVATION(S).

          GREY
FREQUENCY LEVEL
--------- -----
```

FIG. 4.4a Histogram of the data set in the spectral range 0.6 to 0.7
μm (red).

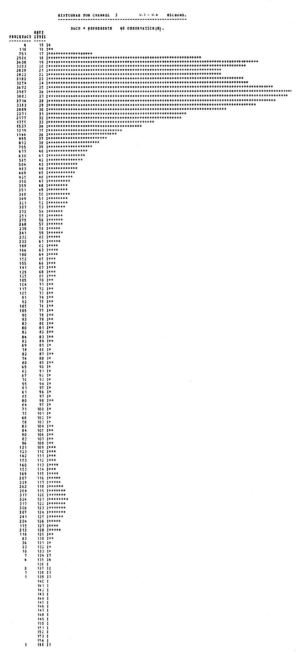

FIG. 4.4b Histogram of the data set in the spectral range 0.5 to 0.6 μm (green).

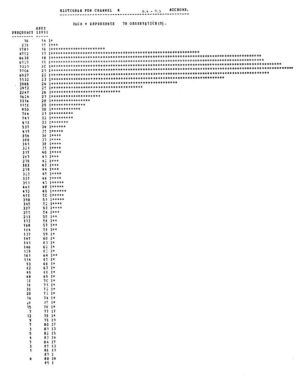

FIG. 4.4c Histogram of the data set in the spectral range 0.4 to 0.5 μm (blue).

ground truth, as we usually call them) can be used to train the system, so that the pattern points in the whole image (254 × 606 pattern points altogether in our data set) can be classified. Figures 4.4(a), (b) and (c) show, respectively, the histograms for the red, green, and blue bands of the aerial photograph.

A portion of the symbol-coding map reproduced by ordinary line printer is shown in Fig. 4.5, in which x's represent object points with gray levels below 20.49; +'s represent those below 26.94 but above 20.49; −'s represent those below 35.48 but above 26.94; and blank's represent those above 35.48. All values are converted to a percentage of the whole gray-level range. A digitized image of the aerial photograph is shown in Fig. 4.6 to check the effectiveness of the use of the line printer. This image was plotted with the norm values of the pixels.

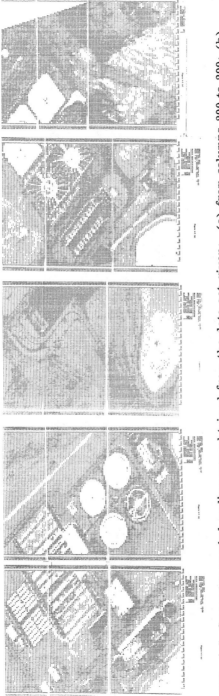

FIG. 4.5 Computer symbol-coding map obtained from the data set given: (a) from columns 200 to 299; (b) from columns 300 to 399; (c) from columns 400 to 499; (d) from columns 500 to 599; (e) from columns 600 to 699.

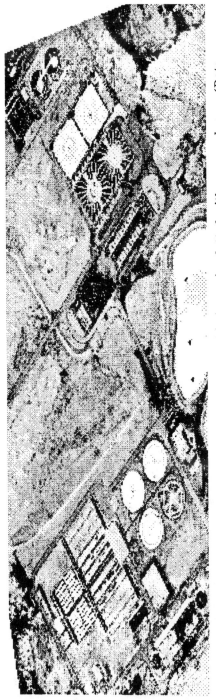

FIG. 4.6 Computer image plot from the data set given. This image is plotted with norm values. (Data courtesy of Frederick Luce, ORSER, Pennsylvania State University.)

PROBLEMS

4.1 We are given the following patterns and their class belongings:

$$(1,0), \ (3,0), \ (2,1), \ (1,2), \ (3,2) \in \omega_1$$

$$(4,3), \ (4,5), \ (5,4), \ (6,3), \ (6,5) \in \omega_2$$

Obtain the equation of the Bayes decision boundary between the two classes by assuming that $p(\omega_1) = p(\omega_2) = 1/2$.

4.2 Consider the following patterns

$$\left.\begin{array}{l} (0,1,0), \ (0,1,1), \ (1,0,1), \ (1,1,1) \\ (0,0,0), \ (0,0,1), \ (1,0,0), \ (1,1,1) \end{array}\right\} \in \omega_1$$

and

$$\left.\begin{array}{l} (5,5,5), \ (5,6,6), \ (6,5,5), \ (6,6,6) \\ (5,6,5), \ (5,5,6), \ (6,5,6), \ (6,6,5) \end{array}\right\} \in \omega_2$$

Find the Bayes decision surface between the two classes of patterns, assuming that they have normal probability density functions and that $p(\omega_1) = p(\omega_2) = 1/2$.

4.3 Derive a Bayes discriminant function for a negative loss function, such as

$$L_{ik} = \begin{cases} -h_i & \text{if } k = i \\ 0 & \text{otherwise } (k \neq i) \end{cases}$$

4.4 Derive the Bayes discriminant function for patterns with independent binary components, and see whether the discriminant function is linear.

5
Clustering Analysis
and Nonsupervised Learning

5.1 INTRODUCTION

5.1.1 Definition of Clustering

What we have discussed so far has been supervised learning; that is, there is a supervisor to teach the system first, how to classify a known set of patterns, and then to let the system go ahead freely classifying other patterns. In such systems we usually need a priori information (information on syntax, semantics, or pragmatics) to form the basis of teaching.

In this chapter we discuss nonsupervised learning, in which the classification process will not depend on a priori information. As a matter of fact, it happens quite frequently that there does not exist much a priori knowledge about the patterns; neither can the proper training pattern sets be obtained.

Clustering is the nonsupervised classification of objects. It is the process of generating classes without any a priori knowledge of prototype classification.

When we are given M patterns, x_1, x_2, ..., x_M, contained in the pattern space S, the process of clustering can be formally stated as: to seek the regions S_1, S_2, ..., S_K such that every x_i, i = 1, 2, ..., M, falls into one of these regions and no x_i falls in two regions; that is,

$$S_1 \cup S_2 \cup S_3 \cdots \cup S_K = S \qquad (5.1)$$

$$S_i \cap S_j = \emptyset \quad \forall \quad i \neq j$$

where $\cup$ and $\cap$ stand for union and intersection, respectively.

Algorithms derived for clustering classify objects into clusters by natural association according to some similarity measures. It is expected that the degree of natural association is high among members belonging to the same category and low among members of different categories.

5.1.2 Similarity Measure

From the definition of clustering, we are to cluster, or form into each class, those patterns $\underline{x}_i$ that are as much alike as possible, and hence we need some kind of similarity measure (or dissimilarity measure). If ζ denotes the dissimilarity measure between two patterns, it is obvious that

$$\zeta(\underline{x}_i, \underline{x}_i) = 0$$

but

$$\zeta(\underline{x}_i, \underline{x}_j) \neq 0 \quad \forall \quad j \neq i \tag{5.2}$$

The similarity measure (or dissimilarity measure) is usually given in numerical form to indicate the degree of natural association or degree of resemblance between patterns in a group, between a pattern and a group of patterns, or between pattern groups.

Many different functions, such as the inertia function and the fuzzy membership function, have also been suggested as the similarity measure, but the most common ones are described next.

Euclidean distance. This is the simplest and most frequently used measure and is represented by

$$d^2(\underline{x}_i, \underline{x}_j) = (\underline{x}_i - \underline{x}_j)^T(\underline{x}_i - \underline{x}_j) = |\underline{x}_i - \underline{x}_j|^2 \tag{5.3}$$

in multidimensional euclidean space. It may be all right to use this distance as a similarity measure if the relative size of the dimension has significance. If not, we should consider weighted euclidean distance, which is

$$d^2(\underline{x}_i, \underline{x}_j) = \sum_{k=1}^{n} \alpha_k (x_{ki} - x_{kj})^2 \tag{5.4}$$

where $\underline{x}_i = [x_{1i}, x_{2i}, \ldots, x_{ni}]^T$; x_{ki} and x_{kj} are the kth components of $\underline{x}_i$ and $\underline{x}_j$, respectively; and α_k is the weighting coefficient. In particular, let us let $\underline{m}_m = [m_{1m}, m_{2m}, \ldots, m_{nm}]^T$ be the mean of the mth cluster (we still presume the class is unknown), and let

$$\alpha_k = \frac{1}{\sigma_{km}^2} \qquad (5.5)$$

where $\underline{\sigma}_m = [\sigma_{1m}, \sigma_{2m}, \ldots, \sigma_{nm}]$ and σ_{km}^2 is the variance of the mth cluster in the kth direction. Then the weighted euclidean distance from $\underline{x}_i$ to the mth cluster is

$$d_m^2(\underline{x}_i, \underline{m}_m) = \sum_{k=1}^{n} \frac{(x_{ki} - m_{km})^2}{\sigma_{km}^2} \qquad (5.6)$$

The cluster shapes obtained by using this measure have loci of equal d_m^2, which are hyperellipsoids aligned with the axes of the n-dimensional pattern space.

Mahalanobis distance. The squared Mahalanobis distance from $\underline{x}_i$ to $\underline{x}_j$ is in the form

$$r^2(\underline{x}_i, \underline{x}_j) = (\underline{x}_i - \underline{x}_j)^T C^{-1} (\underline{x}_i - \underline{x}_j) \qquad (5.7)$$

where C^{-1} is the inverse of the covariance matrix.

Tanimoto coefficient. Tanimoto suggested a similarity ratio known as the Tanimoto coefficient:

$$d_t(\underline{x}_i, \underline{x}_j) = \frac{\underline{x}_i^T \underline{x}_j}{\underline{x}_i^T \underline{x}_i + \underline{x}_j^T \underline{x}_j - \underline{x}_i^T \underline{x}_j} \qquad (5.8)$$

where $\underline{x}_i^T \underline{x}_j$ denotes the number of common attributes between $\underline{x}_i$ and $\underline{x}_j$, $\underline{x}_i^T \underline{x}_i$ denotes the number of attributes possessed by $\underline{x}_i$, and $\underline{x}_j^T \underline{x}_j$ denotes the number of attributes possessed by $\underline{x}_j$. The denominator then gives the number of attributes that are in $\underline{x}_i$ or $\underline{x}_j$ but not in both. The entire expression will therefore represent the ratio of the number of common attributes between $\underline{x}_i$ and $\underline{x}_j$ to the number of attributes that are in either one of the vectors $\underline{x}_i$, $\underline{x}_j$ but not in both.

5.1.3 Types of Clustering Algorithms

Classification of Clustering Algorithms

Lots of clustering algorithms have been suggested. They can be grouped into direct (constructive) or indirect (optimization) algorithms according to whether or not a criterion function is used in the clustering process. For a direct approach, sometimes called the *heuristic approach*, it is simply to isolate pattern classes without the necessity of using a criterion function, whereas for an indirect approach we do use a criterion function to optimize the classification.

Very frequently, clustering algorithms can be classified as an agglomerative or a divisive approach according to the clustering process being worked along the "bottom-up" or the "top-down" direction. A clustering algorithm is said to be *agglomerative* if it starts from isolated patterns and coalesces the nearest patterns or groups according to a threshold from the bottom up to form hierarchies.

A clustering algorithm is said to be *divisive* if it starts from a set of patterns and divides along the top-down direction on minimizing or maximizing some estimating function into optimum clusters.

Many programs have been written in each of these algorithms, but quite a lot have tried to take advantage of both and include both divisive and agglomerative approaches in one program. This leads to another classification based on whether the number of classes is known or unknown beforehand. This is the method we have used in this book.

Intraset and Interset Distances*: One Type of Criterion

We mentioned earlier that the degree of natural association is expected to be high among members belonging to the same category, and low among members of different categories. In other words, the intraset distance should be small, whereas the interset distance should be large.

Mathematically, the interset distance between two separate sets is

$$D_{12} = \overline{D^2([\underline{x}_1^i], [\underline{x}_2^j])} \quad i = 1, 2, \ldots, N_1; \ j = 1, 2, \ldots, N_2$$

$$(5.9)$$

or

$$D_{12} = \frac{1}{N_1 N_2} \sum_{i=1}^{N_1} \sum_{j=1}^{N_2} D^2(\underline{x}_1^i, \underline{x}_2^j) \qquad (5.10)$$

which is the average squared distance between points of separate classes. The subscripts 1 and 2 in the pattern sets $[\underline{x}_1^i]$ and $[\underline{x}_2^j]$ represent classes ω_1 and ω_2, respectively, and N_1 and N_2 are the number of pattern samples in classes ω_1 and ω_2, respectively.

The intraset distance for a set of N patterns (all patterns belonging to the same class) can be derived similarly. Since

---

*See Tou and Gonzalez (1974) and Babu (1973) for supplementary reading.

$$D^2(\underline{x}^i, \underline{x}^j) = \sum_{k=1}^{n} (x_k^i - x_k^j)^2 \tag{5.11}$$

the mean squared distance from a fixed $\underline{x}^i$ to $N - 1$ other patterns in the same set is

$$\overline{D^2(\underline{x}^i, [\underline{x}^j])} = \frac{1}{N-1} \sum_{j=1}^{N} \sum_{k=1}^{n} (x_k^i - x_k^j)^2 \tag{5.12}$$

The intraset distance or the average over all N patterns in the set is then

$$D_{ii} = \overline{D^2([\underline{x}^i], [\underline{x}^j])} = \frac{1}{N} \sum_{i=1}^{N} \left[\frac{1}{N-1} \sum_{j=1}^{N} \sum_{k=1}^{n} (x_k^i - x_k^j)^2 \right] \tag{5.13}$$

or

$$D_{ii} = \frac{N}{N-1} \sum_{k=1}^{n} \left[\frac{1}{N^2} \sum_{j=1}^{N} \sum_{i=1}^{N} (x_k^i - x_k^j)^2 \right] \tag{5.14}$$

Expanding the terms inside the brackets,

$$D_{ii} = \overline{D^2([\underline{x}^i], [\underline{x}^j])}$$

$$= \frac{N}{N-1} \sum_{k=1}^{n} \left[\frac{1}{N} \sum_{j=1}^{N} \frac{1}{N} \sum_{i=1}^{N} (x_k^i)^2 - 2 \frac{1}{N} \sum_{i=1}^{N} x_k^i \frac{1}{N} \sum_{j=1}^{N} x_k^j \right.$$

$$\left. + \frac{1}{N} \sum_{i=1}^{N} \frac{1}{N} \sum_{j=1}^{N} (x_k^j)^2 \right]$$

$$= \frac{N}{N-1} \sum_{k=1}^{n} \left[\frac{1}{N} \sum_{j=1}^{N} \overline{(x_k^i)^2} - 2 \overline{x_k^i} \, \overline{x_k^j} + \frac{1}{N} \sum_{i=1}^{N} \overline{(x_k^j)^2} \right] \tag{5.15}$$

Since we are working on the same pattern set,

$$\overline{(x_k^j)^2} = \overline{(x_k^i)^2} \tag{5.16}$$

we have

$$D_{ii} = \overline{D^2([\underline{x}^i],[\underline{x}^j])} = \frac{2N}{N-1} \sum_{k=1}^{n} \left[\overline{(x_k^i)^2} - \overline{(x_k^i)}^2 \right] \qquad (5.17)$$

Note that, by definition, the variance of the kth component of N patterns is given by

$$(\sigma_k)^2 = \frac{1}{N} \sum_{i=1}^{N} (x_k^i - \overline{x_k^i})^2$$

$$= \frac{1}{N} \sum_{i=1}^{N} (x_k^i)^2 - \frac{2}{N} \sum_{i=1}^{N} x_k^i \overline{x_k^i} + \frac{1}{N} \sum_{i=1}^{N} \overline{x_k^i}^2$$

$$= \overline{(x_k^i)^2} - \overline{(x_k^i)}^2 \qquad (5.18)$$

after simplification. Therefore, the intraset distance is now

$$D_{ii} = \overline{D^2([\underline{x}^i],[\underline{x}^j])} = \frac{2N}{N-1} \sum_{k=1}^{n} \sigma_k^2 \qquad (5.19)$$

or

$$D_{ii} = 2 \sum_{k=1}^{n} \sigma_k^{*2} \qquad (5.19a)$$

where

$$(\sigma_k^*)^2 = \frac{N}{N-1} (\sigma_k)^2 \qquad (5.19b)$$

5.1.4 General Remarks

Most clustering algorithms are heuristic. Most papers on clustering present experimental evidence of results to illustrate the effectiveness of their clustering processes. But to our knowledge, no objective quantitative measure of clustering performance is yet available, although a lot of effort has been expended toward that end. We are not quite certain, at least at the moment, how data dependent the results are.

In clustering applications we generally try to locate the modes, that is, to obtain the local maximum of the probability density if the number M of the class is known. When the number of classes is unknown, we usually try to obtain an estimate of the number and location of modes, that is, to find the natural grouping of patterns. Thus we "learn" something about the statistics. For example, the mean and covariance of the data to be analyzed are useful for data preprocessing and training of the minimum distance classifier for multiple classes as well as for on-line adaptive classification in a nonstationary environment. This is because more significant features from the measurement vectors are extracted to realize more efficient and more accurate pattern classification.

5.2 CLUSTERING WITH AN UNKNOWN NUMBER OF CLASSES

5.2.1 Adaptive Sample Set Construction (Heuristic Method)

When the number of classes is unknown, classification by clustering is actually to construct the probability densities from pattern samples. Adaptive sample set construction is one of the approaches commonly used.

The essential point of this algorithm is to build up the clusters by using distance measure. The first cluster can be chosen arbitrarily. Once the cluster is chosen, try to assign pattern samples to it if the distance from a sample to this cluster center is less than a threshold. If not, form a new cluster. When a pattern sample falls in a cluster, the mean and variance of that cluster will be adjusted. Repeat the process until all the pattern samples are assigned. The whole procedure consists of the following steps:

Step 1: Take the first sample as representative of the first cluster:

$$\underline{z}_1 = \underline{x}_1$$

where $\underline{z}_1$ is the first cluster center.

Step 2: Take the next sample and compute its distance (similarity measure) to all the existing clusters (when starting, there is only one cluster).

(a) Assign $\underline{x}$ to $\underline{z}_i$ (the ith cluster) if

$$d_i(\underline{x}, \underline{z}_i) \leqslant \theta\tau \quad 0 \leqslant \theta \leqslant 1 \tag{5.20}$$

where τ is the membership boundary for a specified cluster. Its value is properly set by the designer.

(b) Do not assign $\underline{x}$ to $\underline{z}_i$ if

$$d_i(\underline{x}, \underline{z}_i) > \tau \tag{5.21}$$

(c) No decision will be made on $\underline{x}$ if $\underline{x}$ falls in the "intermediate region" for $\underline{z}_i$, as shown in Fig. 5.1.

Step 3: (a) Each time a new $\underline{x}$ is assigned to $\underline{z}_i$, compute $\underline{z}_i$ $(t + 1)$ and $\underline{C}(t + 1)$ according to the following expressions:

$$\underline{z}_i(t + 1) = \frac{1}{t + 1}[t\underline{z}_i(t) + \underline{x}] \tag{5.22}$$

$$\underline{C}(t + 1) = \frac{1}{t + 1}[t\underline{C}(t) + (\underline{x} - \underline{z}_i(t + 1))^2] \quad \text{for } i = 1, 2, \ldots, M \tag{5.23}$$

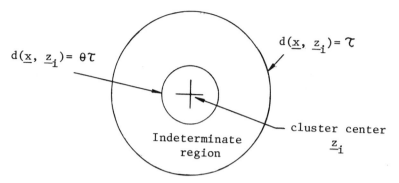

FIG. 5.1 Clustering based on a distance measure.

where t is the number of pattern samples already assigned to z_i and $\underline{x}$ is the (t + 1)th such sample. $z_i(t)$ and $\underline{C}(t)$, the variance, were already computed from the t samples.

    (b) Form a new cluster z_j if

$$d(\underline{x}, \underline{z}_i) > \tau \quad \forall \quad i \qquad (5.24)$$

Step 4: Repeat steps 2 and 3 until all prototypes (pattern samples) have been assigned. There would be some reassignment of $\underline{x}$ when all $\underline{x}$ are again passed through in order. This is because the means and variances have been adjusted with each $\underline{x}$ assigned to z_i.

Step 5: After the training is considered complete (that means that $\underline{x}$ no longer changes class belongings, or some number of $\underline{x}$ are unassigned each time), we can let the system go freely to do the clustering on a large number of pattern samples. No indeterminate region will exist this time. All $\underline{x}$'s falling on the indeterminate region may be assigned to the nearest class according to the minimum distance rule. All those $\underline{x}$'s could be considered unclassified if their distances to all cluster centers are greater than τ.

    This algorithm is simple and efficient. Other than these, it possesses the following advantages: minimum computations are required; pattern samples are processed sequentially without the necessity of being stored; and there is no need to have the number of classes specified.

    On the other hand, there are some drawbacks to the use of this algorithm. First, strong assumptions are required, such as that clusters themselves should be tight and also that clusters should be widely separated from one another. Second, clustering results are dependent on the order of presentation of $\underline{x}$'s and also on the first $\underline{x}$ being used as the initial cluster center. If, for example, cluster center z_i (and also C) changes, or $x(t)$ is presented at a later order $t + m$, that pattern sample might be classified differently. Also, different results of clustering might have resulted during training. Third, clustering results also depend heavily on the value of $\tau\theta$ chosen.

5.2.2 Batchelor and Wilkins' Algorithm

Batchelor and Wilkins suggested another simple heuristic procedure for clustering, sometimes known as the maximin distance algorithm.

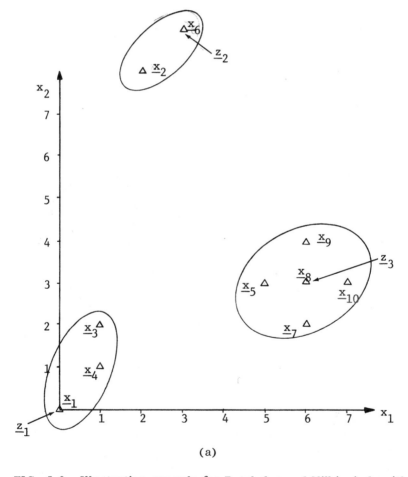

(a)

FIG. 5.2   Illustrative example for Batchelor and Wilkins' algorithm: (a) the 10 two-dimensional patterns. (b) Intermediate clustering results.

1. $z_1 = x_1$

2. $|x_6 - z_1| > |x_2 - z_1| > |x_{10} - z_1| > |x_9 - z_1| > |x_7 - z_1|$
$$> |x_5 - z_1| > |x_3 - z_1| > |x_4 - z_1|$$

   Let $z_2 = x_6$

3. $|x_4 - z_1| < |x_4 - z_2|$

   $|x_3 - z_1| < |x_3 - z_2|$

   $|x_5 - z_1| < |x_5 - z_2|$

   $|x_8 - z_2| = |x_8 - z_1|$

   $|x_7 - z_1| < |x_7 - z_2|$

   $|x_9 - z_2| < |x_9 - z_1|$

   $|x_{10} - z_2| < |x_{10} - z_1|$

   $|x_2 - z_2| < |x_2 - z_1|$

4. Save all distances on left in step 3.

5. The maximum of distances saved in step 4 is $|x_8 - z_2|$.

6. Since $|x_8 - z_2| > \frac{1}{2} |z_2 - z_1|$ let $z_3 = x_8$.

7. $|x_4 - z_1| < |x_4 - z_3| < |x_4 - z_2|$

   $|x_3 - z_1| < |x_3 - z_3| < |x_3 - z_2|$

   $|x_5 - z_3| < |x_5 - z_1| < |x_5 - z_2|$

   $|x_7 - z_3| < |x_7 - z_1| < |x_7 - z_2|$

   $|x_9 - z_3| < |x_9 - z_2| < |x_9 - z_1|$

   $|x_{10} - z_3| < |x_{10} - z_2| < |x_{10} - z_1|$

   $|x_2 - z_2| < |x_2 - z_3| < |x_2 - z_1|$

8. Save all distances on left.

9. The maximum of these minimum distances is $|x_3 - z_1|$.

10. Since $|x_3 - z_1| < \frac{1}{2} \text{Avg}[|z_2 - z_1|, |z_3 - z_2|]$ and since the condition for creation of a new cluster is not satisfied, the algorithm terminates.

(b)

An artificially simple example consisting of 10 two-dimensional patterns
as shown in Fig. 5.2a is used to illustrate the procedure of this
algorithm.

Step 1: Arbitrarily, let $\underline{x}_1$ be the first cluster center, designated
by $\underline{z}_1$.

Step 2: Determine the pattern sample farthest from $\underline{x}_1$, which
is $\underline{x}_6$. Call it cluster center $\underline{z}_2$.

Step 3: Compute the distance from each remaining pattern
sample to $\underline{z}_1$ and $\underline{z}_2$.

Step 4: Save the minimum distance for each pair of these computations.

Step 5: Select the maximum of these minimum distances.

Step 6: If this distance is appreciably greater than a fraction of
the distance $d(\underline{z}_1,\underline{z}_2)$, call the corresponding sample cluster center
$\underline{z}_3$. Otherwise, the algorithm is terminated.

Step 7: Compute the distance from each of the three established
cluster centers to the remaining samples and save the minimum of every
group of three distances. Again, select the maximum of these minimum distances. If this distance is an appreciable fraction of the
"typical" previous maximum distances, the corresponding sample becomes cluster center $\underline{z}_4$. Otherwise, the algorithm is terminated.

Step 8: Repeat until the new maximum distance at a particular
step fails to satisfy the condition for the creation of a new cluster
center.

Step 9: Assign each sample to its nearest cluster center.

Figure 5.2b tabulates the intermediate clustering results. $\underline{x}_1$,
$\underline{x}_6$, and $\underline{x}_8$ are the three cluster centers. $[\underline{x}_1,\underline{x}_3,\underline{x}_4]$, $[\underline{x}_2,\underline{x}_6]$, and
$[\underline{x}_5,\underline{x}_7,\underline{x}_8,\underline{x}_9,\underline{x}_{10}]$ are the three cluster domains.

5.2.3 Hierarchical Clustering Algorithm Based on k-Nearest Neighbors

The process of classifying a point as a member of the class to which its
nearest neighbor belongs is known as *nearest neighbor classification*.
If membership is decided by a majority vote of the k-nearest neighbors, the procedure will be called a k-nearest neighbor decision rule.
The clustering algorithm discussed in this section follows the concept
suggested by Mizoguchi and Kakusho (1978). The procedure consists
of two stages. In the first stage, pregrouping of data is made to obtain subclusters. In the second stage, the subclusters are merged
hierarchically by using a similarity measure.

The procedure in carrying out this algorithm is as follows:

1. Determine k appropriately.
2. Compute $\Omega_k(i)$, $P_k(i)$, and $\xi_k(i)$ for every pattern sample
i, where $\Omega_k(i)$ is a set of k-nearest neighbors of the sample point

i, $i = 1, 2, \ldots, N$, based on euclidean distance measure. $P_k(i)$ is the potential of the pattern sample point i, and is defined as

$$P_k(i) = \frac{1}{k} \sum_{j \in \Omega_k(i)} d(i,j) \qquad (5.25)$$

with $d(i,j)$ the euclidean distance between sample points i and j. Obviously, $d(i,i) = 0$. $\xi_k(i)$ is a set of sample points k-adjacent to the sample point i.

3. Subordinate every point i to the point j such that

$$P_k(j) = \min_{m \in \xi_k(i)} P_k(m) \qquad (5.26)$$

that is, subordinate every sample point i to one of its neighbors j which has the smallest value of potential.

4. Detect and count the subclustering points and assign every point to its nearest subcluster.

5. Merge the subclusters according to whether or not there is a k-boundary point set between the two subclusters.

(a) If there is a k-boundary point set between a pair of subclusters, merge the two most similar subclusters among those unordered pairs of subclusters by a similarity measure, $SIM(m,n)$, which is

$$SIM(m,n) = SIM_1(m,n) * SIM_2(m,n) \qquad (5.27)$$

where m and n denote the subclusters. $SIM_1(m,n)$ represents the difference in density between the cluster and the boundary and can be used to detect the valley of density. $SIM_2(m,n)$ represents the relative size of the boundary to that of the cluster and can be used to detect the neck between two clusters. Mathematically,

$$SIM_1(m,n) = \frac{\min[P_k^{sc}(m), P_k^{sc}(n)]}{\max[BP_k^{m,n}, BP_k^{n,m}]} \qquad (5.28)$$

and

$$SIM_2(m,n) = \frac{N(Y_k^{m,n}) + N(Y_k^{n,m})}{2\min[N(W_m), N(W_n)]} \qquad (5.29)$$

where

$$BP_k^{m,n} = \frac{1}{N(Y_k^{m,n})} \sum_{i \in Y_k^{m,n}} P_k(i) \qquad (5.30)$$

which is the average of $P_k(i)$ over all i in $Y_k^{m,n}$ and

$$P_k^{sc}(m) = \frac{1}{N(\Omega_k(i_m) \cap W_m)} \sum_{i \in \Omega_k(i_m) \cap W_m} P_k(i) \qquad (5.31)$$

which is the average of $P_k(i)$ over all points in cluster m. $N(\cdot)$ denotes the number of elements of the set in parentheses.

$$Y_k^{m,n} = \{i \,|\, i \in W_m \text{ and } \xi_k(i) \cap W_n \neq \emptyset\} \qquad (5.32)$$

denotes the set of points that are in cluster m at the same time that
their respective k adjacent points are in cluster n. W_m is a set of
points contained in subcluster m.

(b) if no k-boundary point set exists between any pair of subclusters, use distance measure to merge subclusters p and q such that

$$d_{min}(p,q) = \min_{(m,n) \in \varphi} d_{min}(m,n) \qquad (5.33)$$

where φ denotes a set of unordered pairs of subclusters.

5.3 CLUSTERING WITH A KNOWN NUMBER OF CLASSES

In this section we assume that the number of classes in the image is
known or that at least a rough idea of the number and locations of
clusters is available. Several algorithms will be introduced.

5.3.1 Minimization of Sum of Squared Distance

This algorithm is based on the minimization of the sum of squared distances from all points in a cluster domain to the cluster center, that
is,

$$\min \sum_{x \in S_j(k)} (\underline{x} - \underline{z}_j)^2 \qquad (5.34)$$

where $S_j(k)$ is the cluster domain for cluster center z_j at the kth iteration. The clustering procedure of this algorithm can be illustrated by means of an example as shown on Fig. 5.3. For clarity 20 two-dimensional pattern samples are considered in this example.

Step 1: Arbitrarily choose two samples as the initial cluster centers.

$$z_1(1) = x_1 = (0,0)^T$$

$$z_2(1) = x_2 = (1,0)^T$$

The number inside the brackets indexes the iteration order.

Step 2: Distribute the pattern samples x among the chosen cluster domains according to the following rule:

$$x_1 \in S_1(1) \quad \text{since} \quad |x_1 - z_1(1)| < |x_1 - z_i(1)|$$

$$\forall i \quad i = 1, 2, \ldots, K \quad i \neq 1 \quad\quad (5.36)$$

$$x_4 \in S_2(1) \quad \text{since} \quad |x_4 - z_2(1)| < |x_4 - z_i(1)|$$

$$\forall i \quad i = 1, 2, \ldots, K \quad i \neq 2 \quad\quad (5.37)$$

etc.

where $K = 2$ in this case. Therefore,

$$S_1(1) = [x_1, x_3]$$

$$S_2(1) = [x_2, x_4, x_5, \ldots, x_{20}]$$

Step 3: Update the cluster centers with Eq. (5.38)

$$z_j(k + 1) = \frac{1}{N_j} \sum_{x \in S_j(k)} x \quad\quad j = 1.2, \ldots, K \quad\quad (5.38)$$

or

$$z_1(2) = \frac{1}{N_1} \sum_{x \in S_1(1)} x = \frac{1}{2}(x_1 + x_3)$$

$$= \begin{pmatrix} 0.0 \\ 0.5 \end{pmatrix}$$

and

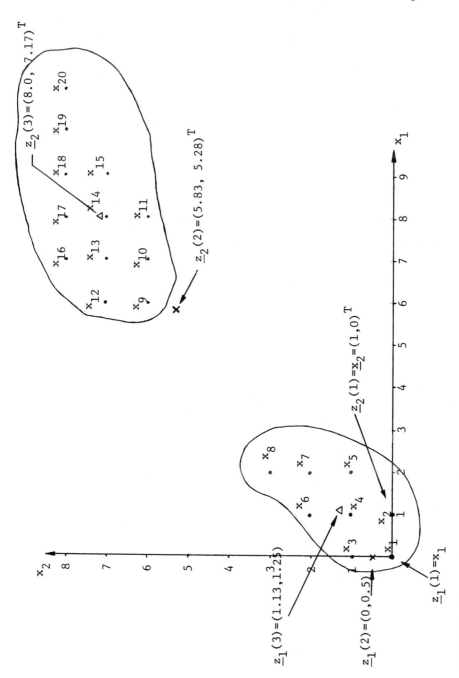

FIG. 5.3 Illustrative example for the K-means algorithm.

and

$$\underline{z}_2(2) = \frac{1}{N_2} \sum_{\underline{x} \in S_2(1)} \underline{x} = \frac{1}{18} (\underline{x}_2 + \underline{x}_4 + \cdots + \underline{x}_{20})$$

$$= \begin{pmatrix} 5.83 \\ 5.28 \end{pmatrix}$$

Note that these adjusted cluster centers, which are the means of all the pattern samples in their respective cluster domains, will minimize the sum of squared distances from all points in $S_j(k)$ to the new cluster centers. In this case, $j = 2$, $k = 2$. N_1 and N_2 are respectively the number of samples in $S_1(1)$ and $S_2(1)$.

  Step 4: Since $\underline{z}_j(2) \neq \underline{z}_j(1)$, $j = 1$, 2, the algorithm has not converged. We return to Step 2 and repeat the process. Otherwise, the procedure is terminated.

  Step 2: with the new cluster centers, we obtain

$$|\underline{x}_\ell - \underline{z}_1(2)| < |\underline{x}_\ell - \underline{z}_2(2)| \qquad \text{for } \ell = 1, 2, \ldots, 8$$

and

$$|\underline{x}_\ell - \underline{z}_2(2)| < |\underline{x}_\ell - \underline{z}_1(2)| \qquad \text{for } \ell = 9, 10, \ldots, 20$$

Cluster domains $S_1(2)$ and $S_2(2)$ are, respectively,

$$S_1(2) = [\underline{x}_1, \underline{x}_2, \ldots, \underline{x}_8]$$

$$S_2(2) = [\underline{x}_9, \underline{x}_{10}, \ldots, \underline{x}_{20}]$$

  Step 3: Update the cluster centers:

$$\underline{z}_1(3) = \frac{1}{N_1} \sum_{\underline{x} \in S_1(2)} \underline{x} = \frac{1}{8} (\underline{x}_1 + \underline{x}_2 + \cdots + \underline{x}_8)$$

$$= \begin{pmatrix} 1.13 \\ 1.25 \end{pmatrix}$$

$$\underline{z}_2(3) = \frac{1}{N_2} \sum_{x \in S_2(2)} \underline{x} = \frac{1}{12} (\underline{x}_9 + \underline{x}_{10} + \cdots + \underline{x}_{20})$$

$$= \begin{pmatrix} 8.0 \\ 7.17 \end{pmatrix}$$

Step 4: Return to step 2, since

$$\underline{z}_j(3) \neq \underline{z}_j(2) \qquad j = 1, 2$$

Step 2: Yields the same results as in the previous iteration:

$$S_1(3) = S_1(2) \quad \text{and} \quad S_2(3) = S_2(2)$$

Step 3: Yields the same results as in the previous iteration.

Step 4: Since $\underline{z}_j(4) = \underline{z}_j(3)$, $j = 1, 2$, the algorithm has converged. The cluster centers we finally obtain are

$$\underline{z}_1 = \begin{pmatrix} 1.13 \\ 1.25 \end{pmatrix}$$

$$\underline{z}_2 = \begin{pmatrix} 8.00 \\ 7.17 \end{pmatrix}$$

From the procedure listed above it is not difficult to see that the cluster centers are sequentially updated. This is why the term "K-means" is sometimes used for this algorithm. It is also not difficult to see that the performance of this K-means algorithm is influenced by the number of cluster centers initially chosen and also by the order in which pattern samples are passed through to the system. It is also influenced by the geometrical properties of the data to be analyzed.

5.3.2 ISODATA Algorithm

ISODATA is an acronym for *I*terative *S*elf-*O*rganizing *D*ata *A*nalysis *T*echniques *A* (the *A* being added to make the word pronounceable). In this algorithm, several process parameters are to be specified:

M = number of clusters desired
η = minimum number of samples desired in a cluster

σ_S = maximum standard deviation allowed in our problem
δ = minimum distance required between clusters
L = maximum number of pairs of cluster centers that can be lumped
I = number of iterations allowed

The procedure of this algorithm can be generalized as follows:

1. Choose some initial cluster centers.
2. Assign patterns to their nearest cluster centers.
3. Recompute the cluster centers (take the average of the samples in their domains as their new cluster centers).
4. Check and see if any cluster does not have enough members. If so, discard that cluster.
5. Compute the standard deviation for each cluster domain and see if it is greater than the maximum value allowed. If so, and if it is also found that the average distance of the samples, in cluster domain S_j from their corresponding cluster center is greater than the overall average distance of the samples from their respective cluster centers, then split that cluster into two.
6. Compute the pairwise distances among all cluster centers. If some of them are smaller than the minimum distance allowed, combine that pair of clusters into one according to some suggested rule.

The whole procedure can be depicted with a flow diagram as shown in Fig. 5.4. Explanations for the terms used in the figure are as follows:

$\underline{x}$ = pattern samples
S_i = cluster domain
$\underline{z}_j$ = cluster center
N_i = number of samples in S_i
N_c = arbitrarily chosen initial number of cluster centers
N = total number of samples
$\overline{D}_i$ = average distance of samples in cluster domain S_i from $\underline{z}_j$
$\overline{D}$ = overall average distance of samples from their respective cluster centers
$\underline{z}_{i\ell}$, $z_{j\ell}$ = cluster centers to be lumped
$N_{i\ell}$, $N_{j\ell}$ = number of samples in clusters $\underline{z}_{i\ell}$, $\underline{z}_{j\ell}$

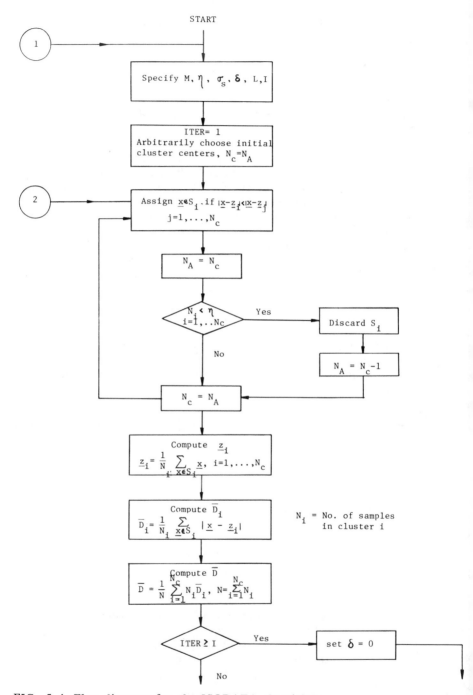

FIG. 5.4 Flow diagram for the ISODATA algorithm.

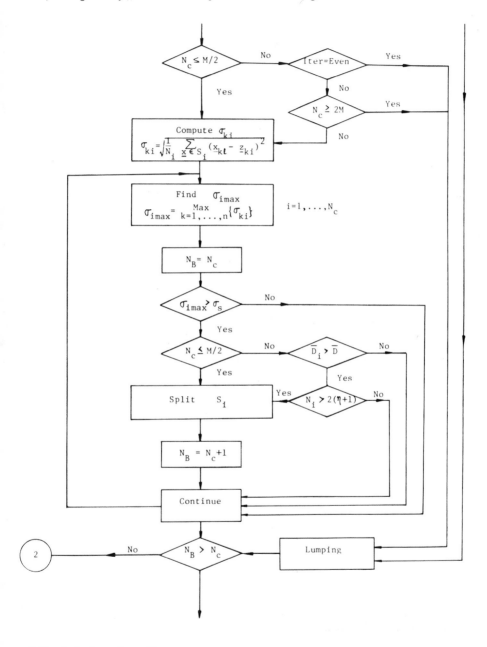

FIG. 5.4 (continued).

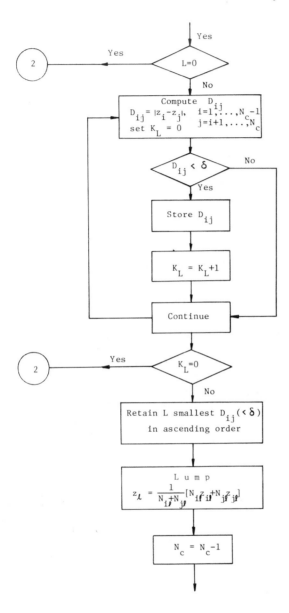

FIG. 5.4 (continued).

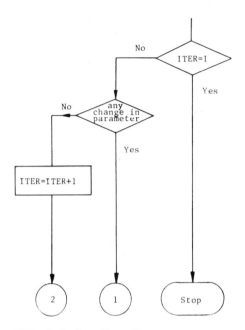

FIG. 5.4 (continued).

N = total number of samples

$\overline{D}_i$ = average distance of samples in cluster domain S_i from z_i

$\overline{D}$ = overall average distance of samples from their respective cluster centers

$\underline{z}_{i\ell}$, $z_{j\ell}$ = cluster centers to be lumped

$N_{i\ell}$, $N_{j\ell}$ = number of samples in clusters $\underline{z}_{i\ell}$, $\underline{z}_{i\ell}$

Example Apply the ISODATA algorithm to the problem of 20 two-dimensional pattern samples discussed in Sec. 5.3.1.

$\underline{x}_1$	$\underline{x}_2$	$\underline{x}_3$	$\underline{x}_4$	$\underline{x}_5$	$\underline{x}_6$	$\underline{x}_7$	$\underline{x}_8$	$\underline{x}_9$	$\underline{x}_{10}$
(0,0)	(1,0)	(0,1)	(1,1)	(2,1)	(1,2)	(2,2)	(2,3)	(6,6)	(7,6)

$\underline{x}_{11}$	$\underline{x}_{12}$	$\underline{x}_{13}$	$\underline{x}_{14}$	$\underline{x}_{15}$	$\underline{x}_{16}$	$\underline{x}_{17}$	$\underline{x}_{18}$	$\underline{x}_{19}$	$\underline{x}_{20}$
(8,6)	(6,7)	(7,7)	(8,7)	(9,7)	(7,8)	(8,8)	(9,8)	(10,8)	(11,8)

Specify the following specified process parameters:

$M = 2$ $\sigma = 4$

$\eta = 1$ $L = 0$

$\sigma_s = 1.5$ $I = 4$

In this case, $N = 20$ and $n = 2$. To start with, let $N_c = 1$, with the initial cluster center being $\underline{z}_1 = (0,0)^T$. Since there is only one cluster center,

$$S_1 = [\underline{x}_1, \underline{x}_2, \ldots, \underline{x}_{20}]$$

and $N_1 = 20$. Since $N_1 > \eta$, no subsets are discarded.
Update the cluster centers:

$$\underline{z}_1' = \frac{1}{20} \sum_{\underline{x} \in S_1} \underline{x} = (5.25, 4.8)^T$$

Compute $\overline{D}_i$:

$$\overline{D}_1 = \frac{1}{N_1} \sum_{\underline{x} \in S_1} |\underline{x} - \underline{z}_1'|$$

$$= 4.42$$

Compute $\overline{D}$. In this case,

$$\overline{D} = \overline{D}_1 = 4.42$$

Since this is not the last iteration and $N = M/2$, find the standard deviation vector:

$$\underline{\sigma}_1 = (\sigma_{11}, \sigma_{21})^T$$

$$\sigma_{11} = \left[\frac{1}{N_1} \sum_{\underline{x} \in S_1} (x_{1k} - z_{11})^2 \right]^{1/2}$$

$$= 3.59$$

$$\sigma_{21} = \left[\frac{1}{N_1} \sum_{\underline{x} \in S_2} (x_{2k} - z_{21})^2 \right]^{1/2}$$

$$= 3.02$$

$$\underline{\sigma}_1 = (3.59, 3.02)^T \qquad \sigma_{1max} = 3.59$$

Since $\sigma_{1max} > \sigma_s$ and $N_c = M/2$, split $\underline{z}_1$. Since $\sigma_{1max} = \sigma_{11}$, split along the first component of $\underline{z}_1$ (let $\gamma = 0.6$, $\gamma\sigma_{1max} = 2.15$).

$$\underline{z}_1'^+ = ((5.25 + 2.15), 4.8)^T = (7.40, 4.8)^T$$

$$\underline{z}_1'^- = ((5.25 - 2.15), 4.8)^T = (3.1, 4.8)^T$$

$$N_c = 2$$

Computing the distance from each sample to the two cluster centers yields the following sample sets:

$$S_1 = [\underline{x}_1, \underline{x}_2, \underline{x}_3, \underline{x}_4, \underline{x}_5, \underline{x}_6, \underline{x}_7, \underline{x}_8]$$

$$S_2 = [\underline{x}_9, \underline{x}_{10}, \underline{x}_{11}, \underline{x}_{12}, \underline{x}_{13}, \underline{x}_{14}, \underline{x}_{15}, \underline{x}_{16}, \underline{x}_{17}, \underline{x}_{18}, \underline{x}_{19}, \underline{x}_{20}]$$

$$N_1 = 8$$

$$N_2 = 12$$

Since both N_1 and N_2 are greater than η, no subsets are discarded. Update the cluster centers:

$$\underline{z}_1 = \frac{1}{N_1} \sum_{\underline{x} \in S_1} \underline{x} = (1.125, 1.25)^T$$

$$\underline{z}_2 = \frac{1}{N_2} \sum_{\underline{x} \in S_2} \underline{x} = (8.00, 7.17)^T$$

Compute $\overline{D}_1$ and $\overline{D}_2$:

$$\overline{D}_1 = \frac{1}{N_1} \sum_{\underline{x} \in S_1} |\underline{x} - \underline{z}_1| = 1.14$$

$$\overline{D}_2 = \frac{1}{N_2} \sum_{\underline{x} \in S_2} |\underline{x} - \underline{z}_2| = 1.49$$

Compute $\overline{D}$:

$$\overline{D} = \frac{1}{N} \sum_{j=1}^{N_c} N_j \overline{D}_j = \frac{1}{20} \sum_{j=1}^{2} N_j \overline{D}_j$$

$$= 1.35$$

This is an even-numbered iteration. Compute D_{12}:

$$D_{12} = |\underline{z}_1 - \underline{z}_2| = 9.07$$

Since $L = 0$, no action is taken. Since the requested number of clusters is satisfied, and the distance between the clusters are large relative to the standard deviations, there is no need to change the parameters.

Distribute the samples again to the two-cluster center according to distance measure. The same results are obtained.

$$\underline{\sigma}_1 = (\sigma_{11}, \sigma_{21})^T$$

$$\sigma_{11} = \left[\frac{1}{N_1} \sum_{\underline{x} \in S_1} (x_{1k} - z_{11})^2 \right]^{1/2}$$

$$= 0.78$$

$$\sigma_{21} = 0.96$$

$$\underline{\sigma}_1 = (0.78, 0.96)^T$$

$$\sigma_{12} = \left[\frac{1}{N_2} \sum_{\underline{x} \in S_2} (x_{1k} - z_{12})^2 \right]^{1/2}$$

$$= 1.47$$

$$\sigma_{22} = \left[\frac{1}{N_2} \sum_{\underline{x} \in S_2} (x_{2k} - z_{22})^2 \right]^{1/2}$$

$$= 0.8$$

$$\underline{\sigma}_2 = (1.47, 0.8)^T$$

$$\sigma_{1max} = 0.96 < \sigma_s$$

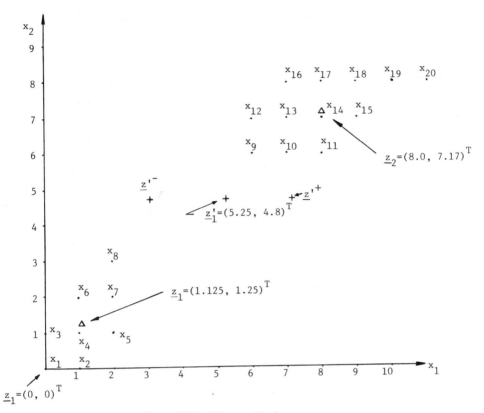

FIG. 5.5 Example of the ISODATA method.

$$\sigma_{2max} = 1.47 < \sigma_s$$

$N_c \geqslant M/2$; therefore, no splitting takes place.
The final results as shown in Fig. 5.5 are

$$z_1 = (1.125, 1.25)^T$$

$$z_2 = (8.00, 7.17)^T$$

$$S_1 = [\underline{x}_1, \underline{x}_2, \underline{x}_3, \underline{x}_4, \underline{x}_5, \underline{x}_6, \underline{x}_7, \underline{x}_8]$$

$$S_2 = [\underline{x}_9, \underline{x}_{10}, \underline{x}_{11}, \underline{x}_{12}, \underline{x}_{13}, \underline{x}_{14}, \underline{x}_{15}, \underline{x}_{16}, \underline{x}_{17}, \underline{x}_{18}, \underline{x}_{19}, \underline{x}_{20}]$$

Note that these results check with those obtained by the Minimization-of-Sum-of-Squared-Distances method discussed in Sec. 5.3.1. But if the σ_s is set at a small value and M is set at higher value, say 3, then z_2 can be further split into two clusters.

5.3.3 Modification of the ISODATA Algorithm (Without Human Intervention in Specifying Certain Process Parameters

As can be seen from Sec. 5.3.2, in using the ISODATA algorithm certain process parameters have to be specified, such as the number of clusters desired, the minimum acceptable standard deviation, and the minimum acceptable distance between clusters. Knowledge of those parameters presumes that previous studies have been done on the data. In addition, the performance of the algorithm is highly dependent on the various parameters preset by the user. The "proper" setting usually can be determined only by a trial-and-error method.

Davies and Bouldin suggested a clustering parameter which is to be minimized to obtain natural partitions of the data sets. The parameter they used is

$$R_{ij} = \frac{D_{ii} + D_{jj}}{D_{ij}} \tag{5.39}$$

where D_{ii} and D_{jj} are defined as the dispersions for clusters i and j, respectively; and D_{ij} is the distance between clusters i and j. It is obvious from the definitions that if $\underline{x}_1, \underline{x}_2, \ldots, \underline{x}_N \in S_p$, the cluster domain, then

$$D_{ii}(\underline{x}_1, \underline{x}_2, \ldots, \underline{x}_N) \geqslant 0 \tag{5.40}$$

$$D_{ii}(\underline{x}_1, \underline{x}_2, \ldots, \underline{x}_N) = 0 \quad \text{iff} \quad \underline{x}_i = \underline{x}_j \quad \forall \ \underline{x}_i, \underline{x}_j \in S_p \tag{5.41}$$

Some limitations on R to make it meaningful are:

$$R_{ij}(D_{ii}, D_{jj}, D_{ij}) \geqslant 0$$

$$R_{ij}(D_{ii}, D_{jj}, D_{ij}) = R_{ji}(D_{jj}, D_{ii}, D_{ji})$$

$$R_{ij}(D_{ii}, D_{jj}, D_{ij}) = 0 \quad \text{iff} \quad D_{ii} = D_{jj} = 0$$

$$\text{If } D_{jj} = D_{kk} \quad \text{and} \quad D_{ij} < D_{ik}$$

then

$$R_{ij}(D_{ii}, D_{jj}, D_{ij}) > R_{ik}(D_{ii}, D_{kk}, D_{ik})$$

$$\text{If } D_{ij} = D_{ik} \quad \text{and} \quad D_{jj} > D_{kk}, \text{ then}$$

$$R_{ij}(D_{ii}, D_{jj}, D_{ij}) > R_{ik}(D_{ii}, D_{kk}, D_{ik})$$

$$\tag{5.42}$$

The first and second expressions indicate that the similarity function R is nonnegative and possesses the property of symmetry. The third expression implies that the similarity between clusters is zero if and only if their dispersions equal zero.

The fourth and fifth expressions imply that if the interset distance between clusters increases while their dispersions remain constant, the similarity of the clusters decreases. On the contrary, if the interest distance remains constant, the similarity of clusters increases when the dispersions increase.

To start with, the initial number of clusters can be chosen as a large number; it can even be chosen as large as the number of patterns in a given data set. By following the ISODATA algorithm as discussed in Sec. 5.3.2, we can obtain the number and locations of clusters as well as the pattern samples allocated to each cluster. D_{ii} and D_{jj} can be computed according to the definition as

$$D_{ii} = \left[\frac{1}{N_i} \sum_{j=1}^{N_i} |\underline{x}_j - \underline{z}_i|^2 \right]^{1/2} \tag{5.43}$$

and

$$D_{ij} = \left[\sum_{k=1}^{n} |z_{ki} - z_{kj}|^2 \right]^{1/2} \tag{5.44}$$

where z_{ki} is the kth component of cluster i. R_{ij} and $\overline{R}$, the average of the similarity measures of each cluster with its most similar cluster, can then be computed according to

$$R_{ij} = \frac{D_{ii} + D_{jj}}{D_{ij}} \tag{5.45}$$

and

$$\overline{R} = \frac{1}{N_c} \sum_{i=1}^{N_c} R_i \tag{5.46}$$

where R_i is the maximum of R_{ij}, $i \neq j$, and N_c is the number of clusters.

Different values of $\overline{R}$ can be obtained for different numbers of clusters in the course of clustering computations. The number of clusters corresponding to the smallest values of $\overline{R}$ seems to be the most appropriate number of clusters.

Figure 5.6a and c show, respectively, a data set of 225 points for test and performance of $\overline{R}$ for the smallest 20 values of N_c. $\overline{R}$ is minimal when $N_c = 8$ and about 5% greater than the minimum when $N_c = 9$.

Figure 5.6b shows that the data points are grouped into seven clusters as shown by the partitioning, and into eight as indicated by the additional dashed line.

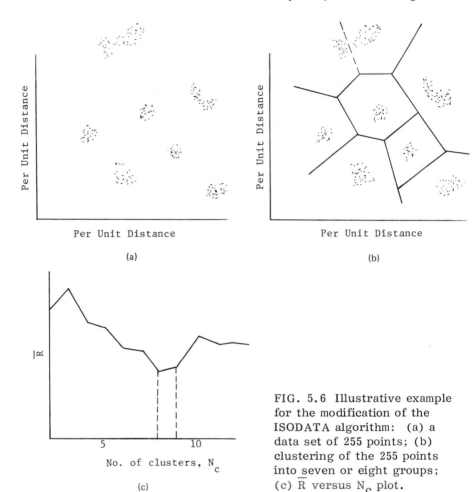

(a)

(b)

No. of clusters, N_c

(c)

FIG. 5.6 Illustrative example
for the modification of the
ISODATA algorithm: (a) a
data set of 255 points; (b)
clustering of the 255 points
into seven or eight groups;
(c) $\bar{R}$ versus N_c plot.

5.3.4 Dynamic Optimal Cluster Seeking Technique (DYNOC)

The DYNOC is an algorithm suggested by Tou to circumvent the short-
comings mentioned earlier. The main point of this algorithm is to
introduce a performance index

$$\lambda(N_c) = \frac{\min\{D_{ij}\}}{\max\{D_{jj}\}} \tag{5.47}$$

to determine the optimal clusters. Optimal clusters occur when $\lambda(N_c)$
reaches a peak and N_c is optimal if $\lambda(N_c)$ is a global maximum. The

maximization of this performance index can be inserted into any of the algorithms, such as the maximum distance algorithm, the K-means algorithm, and the ISODATA algorithm, immediately before splitting of a king size cluster and/or merging of small clusters take place.

5.3.5 Dynamic Clusters Method in Nonhierarchical Clustering

In the methods discussed previously, the cluster center was represented by a simple representative point. In this section we introduce another method, called by Diday the dynamic clusters method, in which a cluster is represented by several representative points called *multicenters* or *sampling*. A Q function is used for determination of the centering. This algorithm can be stated briefly as follows. Given M, the number of clusters, and N_i, the number of pattern points in E_i, i = 1, 2, ..., M, with S = (S_1, ..., S_M) as the M-cluster domains of E and E = (E_1, ..., E_M) as the M sampling of S. Clustering problem is then to find the pair (E,S) that minimizes

$$\Delta(E,S) = \sum_i D(E_i,S_i) = \sum_{\underline{x} \in S_i} \sum_{\underline{z} \in E_i} d(\underline{x},\underline{z}) \tag{5.48}$$

where $E_i \subset E$, i = 1, 2, ..., M, are called sampling or multiple centers or cores; S_i, i = 1, 2, ..., M, are cluster domains with the property $S_i \cap S_j = \emptyset$; $D(E_i,S_i)$ is the "degree of similarity" of E_i to S_i; $d(\underline{x},\underline{z})$, called the *intraset distance*, applies not to a single cluster center, but to a multiple center or core as shown in Fig. 5.7. S = (S_1, S_2, ..., S_M) is achieved such that S_j are formed from the set of elements $\underline{x}$ such that $D(\underline{x},E_i^{(0)}) \leqslant D(\underline{x},E_j^{(0)})$. New samplings $E_i^{(1)}$ can be defined by the N_i elements of E which are closest to S_j in the sense of a certain function Q; that is, the N_i elements of E are chosen such that the following function Q is minimized:

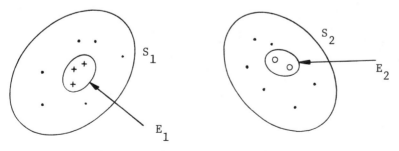

FIG. 5.7 Multicenter representation of a cluster.

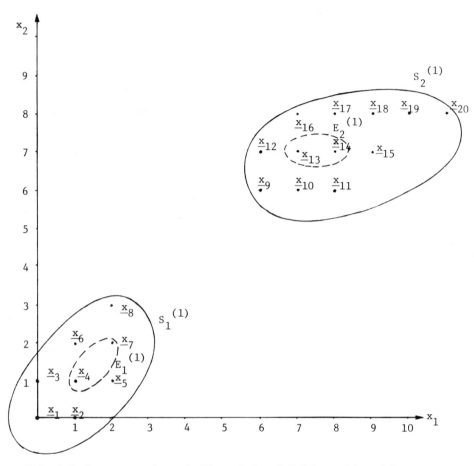

FIG. 5.8 Same example as in Figs. 5.3 and 5.5 but with multicenter representation of a cluster.

$$Q = \frac{D(x, E_i)}{\sum_j D(\underline{x}, E_j)} \qquad (5.49)$$

The choice of the Q function is important in this algorithm. With a good choice of Q, the convergence is generally achieved in about five iterations.

The advantage of this algorithm is that by using N_i multiple centers instead of only the center of gravity, the real form (may be the elongated form) would have been obtainable. If only the center of gravity is used, the recognized form would have been "rounded up."

Use the same example as that used in previous sections to illustrate this algorithm (see Fig. 5.8).

Step 1: Find $S_1^{(0)}$ and $S_2^{(0)}$ by distance measure.

$$S_1^{(0)} = [\underline{x}_1, \underline{x}_2, \underline{x}_3, \underline{x}_4, \underline{x}_5, \underline{x}_6, \underline{x}_7, \underline{x}_9]$$

$$S_2^{(0)} = [\underline{x}_8, \underline{x}_{10}, \underline{x}_{11}, \underline{x}_{12}, \underline{x}_{13}, \underline{x}_{14}, \underline{x}_{15}, \underline{x}_{16}, \underline{x}_{17}, \underline{x}_{18}, \underline{x}_{19}, \underline{x}_{20}]$$

Step 2: Find $E_1^{(1)}$ and $E_2^{(1)}$.

$$E_1^{(1)} = [\underline{x}_4, \underline{x}_7]$$

$$E_2^{(1)} = [\underline{x}_{13}, \underline{x}_{14}]$$

Step 3: Reassign the pattern points to get $S_1^{(1)}$ and $S_2^{(1)}$.

$$S_1^{(1)} = [\underline{x}_1, \underline{x}_2, \underline{x}_3, \underline{x}_4, \underline{x}_5, \underline{x}_6, \underline{x}_7, \underline{x}_8]$$

$$S_2^{(1)} = [\underline{x}_9, \underline{x}_{10}, \underline{x}_{11}, \underline{x}_{12}, \underline{x}_{13}, \underline{x}_{14}, \underline{x}_{15}, \underline{x}_{16}, \underline{x}_{17}, \underline{x}_{18}, \underline{x}_{19}, \underline{x}_{20}]$$

Step 4: Get the correct representation of samplings E_1 and E_2.

5.4 EVALUATION OF CLUSTERING RESULTS BY VARIOUS ALGORITHMS

The principal difficulty in evaluating the results of the various clustering algorithms is the inability to visualize the geometrical properties of a high-dimensional space. However, there are some measures, such as distance between cluster centers, that can be used as a tool for the evaluation of the clustering results.

From the numbers shown on Table 5.1, it can be seen that $\underline{z}_8$ is significantly remote from the other seven cluster centers. Clusters $\underline{z}_1$, $\underline{z}_3$, and $\underline{z}_4$ are close together, as are the clusters $\underline{z}_2$ and $\underline{z}_6$, and $\underline{z}_5$ and $\underline{z}_7$. The number of pattern samples falling into the domain of each cluster is also an aid in interpreting the results. For the example above, if the number of samples associated with the cluster $\underline{z}_8$ is numerous, we will certainly accept it as a cluster center. But when the number of samples is small, cluster $\underline{z}_8$ can be discarded without causing too many discrepancies from the original data.

TABLE 5.1 Distances Between Cluster Centers

Cluster center	$\underline{z}_1$	$\underline{z}_2$	$\underline{z}_3$	$\underline{z}_4$	$\underline{z}_5$	$\underline{z}_6$	$\underline{z}_7$	$\underline{z}_8$
$\underline{z}_1$	0.0	14.8	3.5	2.1	18.0	10.0	21.0	55.6
z_2		0.0	15.4	16.1	23.0	3.5	28.0	54.3
z_3			0.0	5.6	21.0	12.0	19.0	52.8
z_4				0.0	15.0	14.0	19.0	50.0
z_5					0.0	25.0	4.0	49.3
z_6						0.0	23.0	55.8
z_7							0.0	48.2
z_8								0.0

TABLE 5.2 Variances of Various Cluster Domains

Cluster domain	Variance		
	σ_1^2	σ_2^2	σ_3^2
$\underline{z}_1$	1.1	0.8	0.7
$\underline{z}_2$	1.8	1.5	1.0
$\underline{z}_3$	2.5	3.6	5.7
$\underline{z}_4$	2.5	3.8	19.5
$\underline{z}_5$	4.1	4.7	5.4
$\underline{z}_6$	3.7	3.9	5.5
$\underline{z}_7$	4.2	8.6	4.8

Another bit of useful information that can be used in the evaluation of clustering is the variance of each cluster domain about its mean. Variances are useful to infer the relative distribution of the samples in the domains. From the component values of a variance along the coordinate axes, we can estimate the tightness of the pattern points around the cluster as well as the shape of the cluster domain. For the cluster z_1 shown in the variance table (Table 5.2) we can say that it has a hyperspherical shape for its domain, since σ_i^2, $i = 1, 2, 3$, are almost the same for each component. But for cluster z_4, the shape of its cluster domain will be somewhat elongated about the third coordinate axis.

The ratio of interest distance to intraset distance is another criterion for the evaluation of clustering results. We definitely prefer a high performance value for this ratio.

Other quantitative measures of the clustering properties can be the closest and most distant points from the cluster center in each domain and the covariance matrix of each sample set. Computational complexity and computer time are other measures for comparison.

5.5 GRAPH THEORETIC METHODS

The disadvantage of the approaches discussed so far is that the clustering results are dependent on the presentation ordering of the pattern samples. One could argue that the clusters might be determined more accurately if all the samples were considered simultaneously. Graph theoretic approaches are suggested to meet such requirements, but at a possible increase in computational time and also at a substantial cost in rapid-access storage.

5.5.1 Similarity Matrix

The similarity matrix is such a matrix used to show the degree of similarity between a variety of pattern points. Consider that matrix as an $N \times N$ symmetric matrix whose elements are

$$s_{ij} = \begin{cases} 1, & d(\underline{x}_i, \underline{x}_j) \leq \theta \\ 0, & d(\underline{x}_i, \underline{x}_j) > \theta \end{cases} \quad i, j = 1, 2, \ldots, N \qquad (5.50)$$

where $d(\underline{x}_i, \underline{x}_j)$ is the intersample distance between pattern points $\underline{x}_i$ and $\underline{x}_j$. θ is the threshold distance used to denote the similarity between the two pattern points. In other words, s_{ij} tells whether the pair of samples are closer than a distance θ. s_{ij} are binary numbers chosen such that only one bit of storage is required by $d(\underline{x}_i, \underline{x}_j)$.

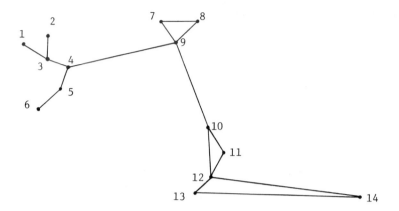

FIG. 5.9 Illustrative example of 14 samples for similarity matrix studies.

Figure 5.9 shows a two-dimensional plot for a set of 14 samples. Its corresponding similarity matrix drawn from the two-dimensional samples above is shown below.

	1	2	3	4	5	6	7	8	9	10	11	12	13	14
1	1	1	1	0	0	0	0	0	0	0	0	0	0	0
2	1	1	1	0	0	0	0	0	0	0	0	0	0	0
3	1	1	1	1	0	0	0	0	0	0	0	0	0	0
4	0	0	0	1	1	0	0	0	0	0	0	0	0	0
5	0	0	0	1	1	1	0	0	0	0	0	0	0	0
6	0	0	0	0	1	1	0	0	0	0	0	0	0	0
7	0	0	0	0	0	0	1	1	1	0	0	0	0	0
8	0	0	0	0	0	0	1	1	1	0	0	0	0	0
9	0	0	0	0	0	0	1	1	1	0	0	0	0	0
10	0	0	0	0	0	0	0	0	0	1	1	0	0	0
11	0	0	0	0	0	0	0	0	0	1	1	1	0	0
12	0	0	0	0	0	0	0	0	0	0	1	1	1	0
13	0	0	0	0	0	0	0	0	0	0	0	1	1	0
14	0	0	0	0	0	0	0	0	0	0	0	0	0	1

$S =$ (matrix above), $\theta = 2$

This similarity matrix can be used for clustering. The procedure is as follows:

1. Choose the row of S with the most 1's (the choice is arbitrary if there is more than one such row), say row i.
2. Form a cluster of $\underline{x}_i$ and all $\underline{x}_j$ corresponding to the 1's in row i.
3. Add $\underline{x}_k$ to the cluster if $s_{jk} = 1$; that is, if $\underline{x}_j$ is already in the cluster and $d(\underline{x}_j,\underline{x}_k) \leq \theta$ (or $s_{jk} = 1$), then $\underline{x}_k$ should also be in the cluster even if $s_{ik} = 0$ [i.e., $d(\underline{x}_i,\underline{x}_k) > \theta$].
4. Repeat step 1 until no new $\underline{x}$'s can be added to the cluster.
5. Remove all columns and rows corresponding to $\underline{x}$'s in the cluster to form a reduced matrix.
6. Repeat steps 1 through 5 for the reduced matrices until no further reductions are possible (i.e., no more clusters can be formed).

For our example, choose $\theta = 2$.

1. Choose row 1 with three 1's in it.
2. $[\underline{x}_1,\underline{x}_2,\underline{x}_3]$ form a cluster $[s_{11} = s_{12} = s_{13} = 1]$.
3-4. Samples $\underline{x}_1$, $\underline{x}_2$, and $\underline{x}_3$ are drawn into the cluster. Row 3 has a 1 in column 4; therefore, $\underline{x}_4$ is added to the cluster, resulting in the cluster $[\underline{x}_1,\underline{x}_2,\underline{x}_3,\underline{x}_4]$. By the same token, $\underline{x}_5$ and $\underline{x}_6$ are also added to the cluster, resulting in a cluster consisting of $[\underline{x}_i]$, $i = 1, 2, \ldots, 6$.
5. The reduced matrix after removal of all columns and rows corresponding to $\underline{x}$'s in the cluster:

	7	8	9	10	11	12	13	14
7	1	1	1	0	0	0	0	0
8	1	1	1	0	0	0	0	0
9	1	1	1	0	0	0	0	0
S' = 10	0	0	0	1	1	0	0	0
11	0	0	0	1	1	1	0	0
12	0	0	0	0	1	1	1	0
13	0	0	0	0	0	1	1	0
14	0	0	0	0	0	0	0	1

1. Choose row 7 with three 1's in that row.
2. $[\underline{x}_7,\underline{x}_8,\underline{x}_9]$ form a cluster.

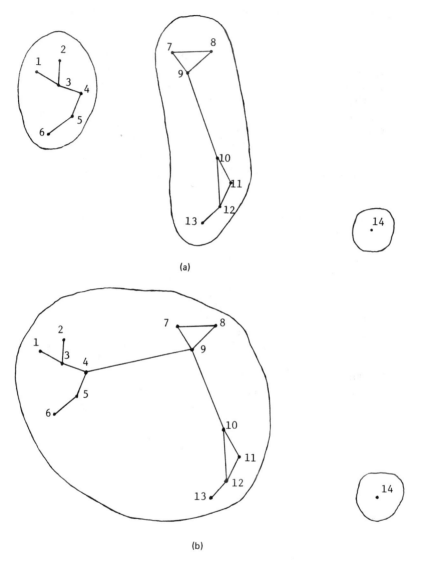

(a)

(b)

FIG. 5.10 Effect of the choice of θ on clustering: (a) θ = 4, three clusters; (b) θ = 5, two clusters; (c) θ = 8, one cluster.

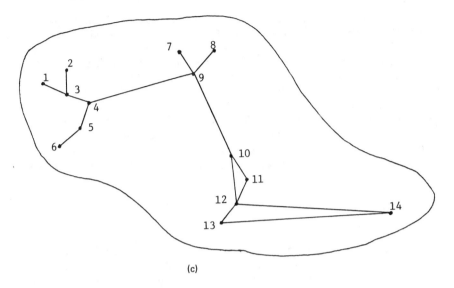

(c)

3-4. Since there are no $s_{jk} = 1$ for $j = 9$, no $\underline{x}_k$ is added to this cluster.

5. The reduced matrix after removal of all columns and rows corresponding to $\underline{x}$'s in the cluster:

$$
\begin{array}{c}
\quad\quad\quad 10 \quad 11 \quad 12 \quad 13 \quad 14 \\[4pt]
\begin{array}{r}
10 \\
11 \\
S'' = 12 \\
13 \\
14
\end{array}
\left|
\begin{array}{ccccc}
1 & 1 & 0 & 0 & 0 \\
1 & 1 & 1 & 0 & 0 \\
0 & 1 & 1 & 1 & 0 \\
0 & 0 & 1 & 1 & 0 \\
0 & 0 & 0 & 0 & 1
\end{array}
\right|
\end{array}
$$

1. Choose row 11 with three 1's in that row.
2. $[\underline{x}_{10}, \underline{x}_{11}, \underline{x}_{12}]$ form a cluster.
3-4. Since $s_{12,13} = 1$, $\underline{x}_{13}$ is added to the cluster according to procedure 3 as stated above. The cluster then consists of $[\underline{x}_{10}, \underline{x}_{11}, \underline{x}_{12}, \underline{x}_{13}]$.
5. The reduced matrix after removal of all columns and rows corresponding to $\underline{x}$'s in the cluster:

$$
\begin{array}{c}
14 \\[2pt]
S''' = 14 \left| \; 1 \; \right|
\end{array}
$$

leaving $[\underline{x}_{14}]$ as the final cluster.

When θ is chosen, clusters are defined by disjoint connected subgraphs of the graph so defined. Obviously, the choice of the value of θ is critical. Assume for the example above that θ = 4; then three clusters will be formed from the same data set. When the θ chosen becomes larger, say θ = 8, there will be only one cluster (see Fig. 5.10c). If θ = 1, there will be 14 clusters with one pattern point in each cluster.

Note that in this method there are a total of N^2 elements in S and $N(N - 1)/2$ nonredundant elements (distances) in S. This may impose a severe limitation on the number of pattern points that can be examined. If we have 1000 samples, these will generate about 500,000 interpoint distances.

5.5.2 Spanning Tree Methods

Minimal Spanning Tree Method

The minimal spanning tree method results from a graph analysis of arbitrary point sets of data. Before our discussion on this method, let us introduce some terms that should prove useful. Given a set G of points $\underline{x}_i$, i = 1, 2, ..., N:

1. An *edge* is a connection between two points.
2. A *path* is a sequence of edges connecting two points.
3. A *loop* is a closed path.
4. A *connected graph* has one or more paths between any pair or points.
5. A *tree* is a connected graph with no closed loops.
6. A *spanning tree* is a tree that contains every point in G.
7. The *weight* of a tree is the sum of weights assigned to each edge in the tree; for example, the weight equals the distance between two points at the ends of the edge.
8. The *minimal spanning tree* (MST) is that spanning tree of minimal weight (among all possible spanning trees of G).
9. The *main diameter* is that path of the MST containing the largest number of points (formed by removing the branch points from the minimal spanning tree).

Figure 5.11 shows the minimal spanning tree and its original data. Because the minimum spanning tree is unique to a set of points in terms of a minimum total weight, it is possible to use the tree as a basis for cluster detection by combining both distance properties and density properties.

Several main diameters can be drawn. Two of them are shown in Figs. 5.12 and 5.13 with their edge weight plots.

From the minimal spanning tree shown in Fig. 5.11 clusterings can be found by the nearest neighbor algorithm. Removal of the longest edge, 12-14, produces a two-cluster grouping; further removal

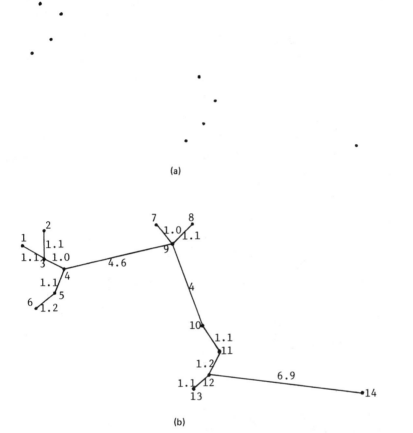

(a)

(b)

FIG. 5.11 Minimal spanning tree: (a) original data; (b) minimal spanning tree.

of the next longest edge, 4-9, produces a three-cluster grouping; and removal of all three long edges produces a four-cluster grouping. These correspond to choosing breaks where maximum weights occur in the main-diameter histogram.

Shared Near Neighbor Maximal Spanning Tree for Clustering

The method by Jarvis that we describe next is one that associates the shared near neighbor rules with the spanning tree in a graph theoretic framework. In this method the influence of other points in the set is taken

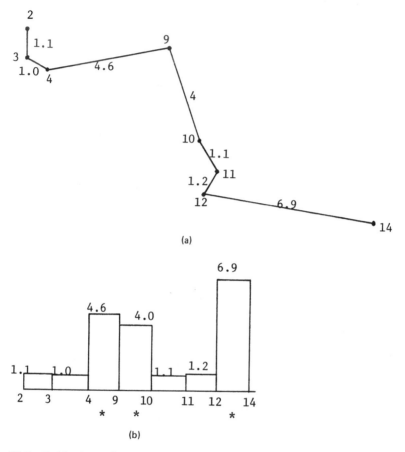

(a)

(b)

FIG. 5.12 One of the main diameters of the MST shown in FIG. 5.11: (a) main diameter; (b) edge weight plot.

into consideration quantitatively on the relative similarity of each pair of pattern points. The idea behind this shared near neighbor max-imal spanning tree concept is to transform context-insensitive measures into ones that reflect an interaction of point placement relationships in the relative vicinity of the candidate pair. In this method it is pre-sumed that pairs of points in the set are similar to the extent that they share the same near neighbors provided that each is in the defined near neighborhood of the other.

The procedure can be stated as follows:

Step 1: List the k-nearest neighbors for each pattern point $\underline{x}_i$, i = 1, 2, ..., N, shown in FIG. 5.14 in order of closeness, as shown

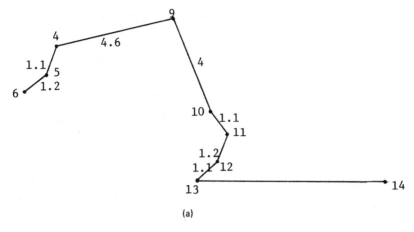

(a)

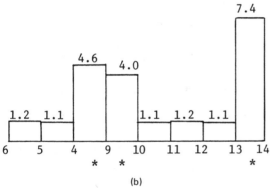

(b)

FIG. 5.13 One of the main diameters of the MST shown in Fig. 5.11: (a) main diameter; (b) edge weight plot.

$\underline{x}_1$ $\quad$ $\underline{x}_2$ $\quad$ $\underline{x}_3$ $\quad$ $\underline{x}_4$ $\quad$ $\underline{x}_5$ $\quad$ $\underline{x}_6$

$\underline{x}_7$ $\quad$ $\underline{x}_8$ $\quad$ $\underline{x}_9$ $\quad$ $\underline{x}_{10}$ $\quad$ $\underline{x}_{11}$ $\quad$ $\underline{x}_{12}$

$\underline{x}_{13}$ $\quad$ $\underline{x}_{14}$ $\quad$ $\underline{x}_{15}$ $\quad$ $\underline{x}_{16}$ $\quad$ $\underline{x}_{17}$ $\quad$ $\underline{x}_{18}$

$\underline{x}_{19}$ $\quad$ $\underline{x}_{20}$ $\quad$ $\underline{x}_{21}$ $\quad$ $\underline{x}_{22}$ $\quad$ $\underline{x}_{23}$ $\quad$ $\underline{x}_{24}$

$\underline{x}_{25}$ $\quad$ $\underline{x}_{26}$ $\quad$ $\underline{x}_{27}$ $\quad$ $\underline{x}_{28}$ $\quad$ $\underline{x}_{29}$ $\quad$ $\underline{x}_{30}$

FIG. 5.14 Simple data set used for illustration of the shared near neighbor maximal spanning tree method.

TABLE 5.3 N × (k + 1) Nearest Neighbor Integer Matrix for the Simple N-point Data Set Shown in Fig. 5.14

Point	0	1	2	3	4	5	6	7	8
$\underline{x}_8$	$\underline{x}_8$	$\underline{x}_2$	$\underline{x}_9$	$\underline{x}_{14}$	$\underline{x}_7$	$\underline{x}_3$	$\underline{x}_{15}$	$\underline{x}_{13}$	$\underline{x}_1$
$\underline{x}_9$	$\underline{x}_9$	$\underline{x}_3$	$\underline{x}_{10}$	$\underline{x}_{15}$	$\underline{x}_8$	$\underline{x}_4$	$\underline{x}_{16}$	$\underline{x}_{14}$	$\underline{x}_2$
$\underline{x}_{10}$	$\underline{x}_{10}$	$\underline{x}_4$	$\underline{x}_{11}$	$\underline{x}_{16}$	$\underline{x}_9$	$\underline{x}_5$	$\underline{x}_{17}$	$\underline{x}_{15}$	$\underline{x}_3$
$\underline{x}_{11}$	$\underline{x}_{11}$	$\underline{x}_5$	$\underline{x}_{12}$	$\underline{x}_{17}$	$\underline{x}_{10}$	$\underline{x}_6$	$\underline{x}_{18}$	$\underline{x}_{16}$	$\underline{x}_4$
$\underline{x}_{14}$	$\underline{x}_{14}$	$\underline{x}_8$	$\underline{x}_{15}$	$\underline{x}_{20}$	$\underline{x}_{13}$	$\underline{x}_9$	$\underline{x}_{21}$	$\underline{x}_{19}$	$\underline{x}_7$
$\underline{x}_{15}$	$\underline{x}_{15}$	$\underline{x}_9$	$\underline{x}_{16}$	$\underline{x}_{21}$	$\underline{x}_{14}$	$\underline{x}_{10}$	$\underline{x}_{22}$	$\underline{x}_{20}$	$\underline{x}_8$
$\underline{x}_{16}$	$\underline{x}_{16}$	$\underline{x}_{10}$	$\underline{x}_{17}$	$\underline{x}_{22}$	$\underline{x}_{15}$	$\underline{x}_{11}$	$\underline{x}_{23}$	$\underline{x}_{21}$	$\underline{x}_9$
$\underline{x}_{17}$	$\underline{x}_{17}$	$\underline{x}_{11}$	$\underline{x}_{18}$	$\underline{x}_{23}$	$\underline{x}_{16}$	$\underline{x}_{12}$	$\underline{x}_{24}$	$\underline{x}_{22}$	$\underline{x}_{10}$
$\underline{x}_{20}$	$\underline{x}_{20}$	$\underline{x}_{14}$	$\underline{x}_{21}$	$\underline{x}_{26}$	$\underline{x}_{19}$	$\underline{x}_{15}$	$\underline{x}_{27}$	$\underline{x}_{25}$	$\underline{x}_{13}$
$\underline{x}_{21}$	$\underline{x}_{21}$	$\underline{x}_{15}$	$\underline{x}_{22}$	$\underline{x}_{27}$	$\underline{x}_{20}$	$\underline{x}_{16}$	$\underline{x}_{28}$	$\underline{x}_{26}$	$\underline{x}_{14}$
$\underline{x}_{22}$	$\underline{x}_{22}$	$\underline{x}_{16}$	$\underline{x}_{23}$	$\underline{x}_{28}$	$\underline{x}_{21}$	$\underline{x}_{17}$	$\underline{x}_{29}$	$\underline{x}_{27}$	$\underline{x}_{15}$
$\underline{x}_{23}$	$\underline{x}_{23}$	$\underline{x}_{17}$	$\underline{x}_{24}$	$\underline{x}_{29}$	$\underline{x}_{22}$	$\underline{x}_{18}$	$\underline{x}_{30}$	$\underline{x}_{28}$	$\underline{x}_{16}$

in Table 5.3. The simplest euclidean distance measure can be used for this purpose. The k-nearest neighbor N × (k + 1) matrix generated is to be used in subsequent processing.

Step 2: Test for occurrence of the first entry of each row in the other rows of the matrix to find pairs of rows for later processing (usually not more than k rows can be found).

Step 3: Count the number of index matches between the two rows. If the number of matches exceeds k_t (a threshold number to be set), the two points indexed in the first column of the two rows are said to be in the same cluster.

Use the match count as a similarity index (see Table 5.4) to develop single-link and MST-like structures against orderings of this new measure. As the near-neighbor-sharing number in this case is a similarity measure (not a distance measure), the structure is a max-

TABLE 5.4 Index Match Count Between Rows in Table 5.3[a]

Point	$\underline{x}_8$	$\underline{x}_9$	$\underline{x}_{10}$	$\underline{x}_{11}$	$\underline{x}_{14}$	$\underline{x}_{15}$	$\underline{x}_{16}$	$\underline{x}_{17}$	$\underline{x}_{20}$	$\underline{x}_{21}$	$\underline{x}_{22}$	$\underline{x}_{23}$
$\underline{x}_8$	8											
$\underline{x}_9$	6	8										
$\underline{x}_{10}$	3	6	8									
$\underline{x}_{11}$	0	3	6	8								
$\underline{x}_{14}$	6	4	2	0	8							
$\underline{x}_{15}$	4	6	4	2	6	8						
$\underline{x}_{16}$	2	4	6	4	3	6	8					
$\underline{x}_{17}$	0	2	4	6	0	3	6	8				
$\underline{x}_{20}$	3	2	1	0	6	4	2	0	8			
$\underline{x}_{21}$	2	3	2	1	4	6	4	2	6	8		
$\underline{x}_{22}$	1	2	3	2	2	4	6	4	3	6	8	
$\underline{x}_{23}$	0	1	2	3	0	2	4	6	0	3	6	8

[a]Boundary points are not included.

imum spanning tree. Use absolute thresholds to cut edges in the maximal spanning tree and define the resulting cluster properties in the
single linkage context.

Figure 5.15 shows another example, consisting of a point set and
its corresponding euclidean metric minimal spanning tree and the
shared near neighbor maximal spanning tree for $k_t = 10$. The links
with the smallest sharing number are marked "I," those with the next
smallest are marked "II," and so on up to "IV." These markings
indicate how a hierarchy of clusters would form.

Graph Theoretic Clustering Based on Limited Neighborhood Sets

Most of the approaches discussed previously were based on distance
measure, which is effective in many applications. But difficulties
would occur if this simple distance measure were employed for the
clustering of certain types of data sets, such as those with a change

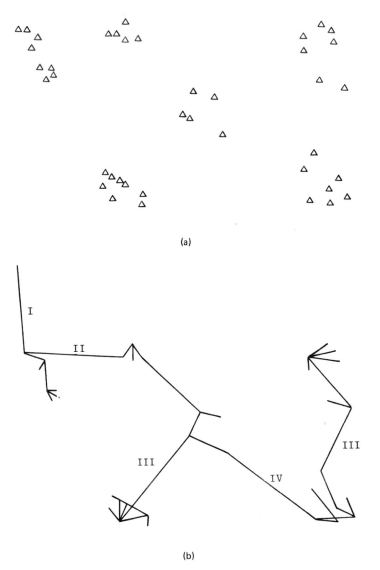

(a)

(b)

FIG. 5.15 Another example: (a) a point set and (b) its shared near neighbor maximal spanning tree for k = 10.

in point density, those with a neck between subclusters, and those with chained clusters within the set, as shown in Fig. 5.16(a), (b), and (c), respectively, where it is not difficult to identify the clusters visually.

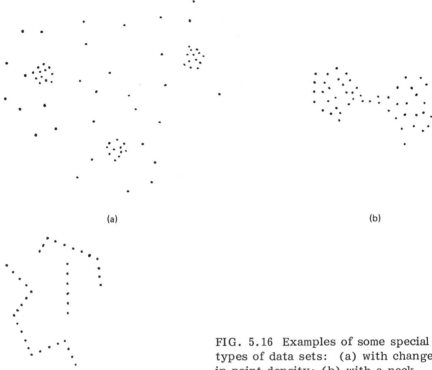

(a)

(b)

(c)

FIG. 5.16 Examples of some special types of data sets: (a) with change in point density; (b) with a neck between subclusters; (c) with chain clusters.

The method discussed in this section is designated primarily for such problems. It is based on the limited neighborhood concept, which originated from the visual perceptual model of clusters.

Several definitions are useful for illustrating this method. Let

$$\mathcal{S} = [S_1, S_2, \ldots, S_M] \quad \text{and} \quad \mathcal{R} = [R_1, R_2, \ldots, R_M]$$

where S_ℓ and R_ℓ, $\ell = 1, 2, \ldots, M$, represent, respectively, the graphs and the regions of influence; (p_i, p_j) represents a graph edge joining points p_i and p_j. To illustrate the region of influence, two graphs should be defined: the Gabriel graph and the relative neighborhood graph.

The *Gabriel graph* (GG) is defined in terms of circular regions. Line segment (p_i, p_j) is included as an edge of the GG if no other point p_k lies within or on the boundary of the circle with (p_i, p_j) as the diameter, as shown in Fig. 5.17a.

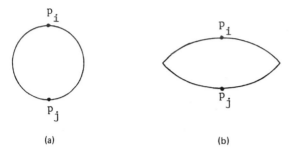

(a)   (b)

FIG. 5.17 Definitions for a Gabriel graph and relative neighborhood graph: (a) circular region defined by GG; (b) lune region defined by RNG.

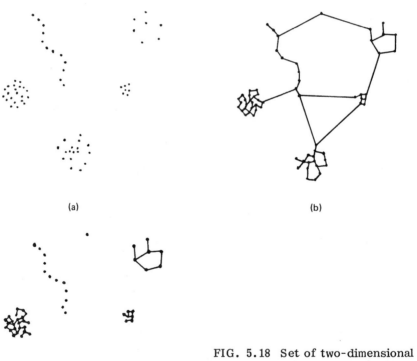

(a)   (b)

(c)

FIG. 5.18 Set of two-dimensional dot patterns to illustrate a graph theoretic clustering algorithm based on limited neighborhood sets: (a) data set; (b) one cluster; (c) six clusters.

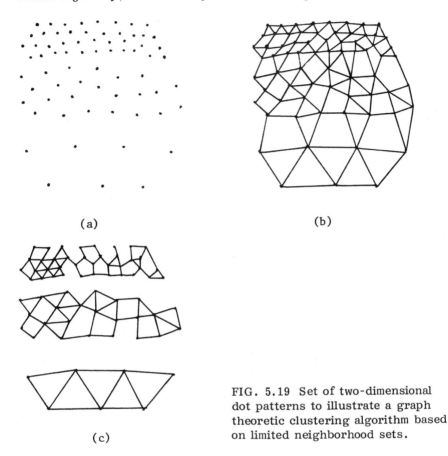

(a) (b)

(c)

FIG. 5.19 Set of two-dimensional dot patterns to illustrate a graph theoretic clustering algorithm based on limited neighborhood sets.

Similarly, the *relative neighborhood graph* (RNG) is defined in terms of a lune region. Line segment (p_i, p_j) is included as an edge of the RNG if no other point p_k lies within or on the boundary of the lune, with p_i and p_j as the two points on the circular arcs of the lunes. S_ℓ and R_ℓ can then be defined as

$$(p_i, p_j) \in S_\ell \quad \text{iff} \quad p_k \notin R_\ell(p_i, p_j)$$

$$\forall \quad k = 1, 2, \ldots, n; \quad k \neq i \neq j \qquad (5.51)$$

$$R_\ell(p_i, p_j) = \{\underline{x}: f[d(\underline{x}, p_i), d(\underline{x}, p_j)] < d(p_i, p_j); \quad i \neq j\} \qquad (5.52)$$

From the two definitions above it can be seen that S_ℓ defines a limited neighborhood set. If $\max[d(\underline{x}, p_i), d(\underline{x}, p_j)]$ is chosen for the function $f[d(\underline{x}, p_i), d(\underline{x}, p_j)]$ in Eq. (5.52) [i.e., we find the maximum between $d(\underline{x}, p_i)$ and $d(\underline{x}, p_j)$ and use it for the f function], then we obtain

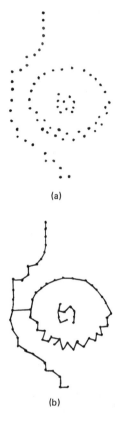

(a)

(b)

FIG. 5.20 Set of two-dimensional dot patterns to illustrate a graph theoretic clustering algorithm based on limited neighborhood sets.

$$R_{RNG}(p_i,p_j) = \{\underline{x}: \max[d(\underline{x},p_i),d(\underline{x},p_j)] < d(p_i,p_j), \ i \neq j\}$$

$$(5.53)$$

where $R_{RNG}(p_i,p_j)$ represents the RNG region of influence. When

$$d^2(\underline{x},p_i) + d^2(\underline{x},p_j)$$

is used for

$$f[d(\underline{x},p_i),d(\underline{x},p_j)]$$

we have

$$R_{GG}(p_i,p_j) = \{\underline{x}: d^2(\underline{x},p_i) + d^2(\underline{x},p_j) \leq d^2(p_i,p_j), \ i \neq j \quad (5.54)$$

where $R_{GG}(p_i,p_j)$ represents the GG region of influence.

The definition of R_ℓ will determine the property of S_ℓ. If $R_\ell \subseteq R_{GG}$, the edges of S_ℓ will be nonintersecting. But if $R_\ell \supset R_{GG}$, intersecting edges are allowed.

Take an example to illustrate this. Assume that we have regions of influence such as the following:

$$R_1(p_i,p_j,\beta) = R_{GG}(p_i,p_j) \cup \{\underline{x}: \beta \min[d(\underline{x},p_i),d(\underline{x},p_j)]$$

$$< d(p_i,p_j), \quad i \ne j\}$$

$$R_2(p_i,p_j,\beta) = R_{RNG}(p_i,p_j) \cup \{\underline{x}: \beta \min[d(\underline{x},p_i),d(\underline{x},p_j)]$$

$$< d(p_i,p_j), \quad i \ne j\}$$

where $0 < \beta < 1$ is a factor of relative edge consistency. Thus $S_1(\beta)$ is obtained from the GG by removing edges (p_i,p_j) if

$$\frac{d(p_i,p_j)}{\min[d(p_i,p_a),d(p_j,p_b)]} > \beta$$

where p_a ($\ne p_j$) denotes the nearest Gabriel neighbor to p_i and p_b ($\ne p_i$) denotes the nearest Gabriel neighbor to p_j.

It is then clear that varying β would control the fragmentation of the data set and hence would give a sequence of nested clusterings. Increasing β would break the data set into a greater number of smaller clusters. The examples of two-dimensional dot patterns shown in Figs. 5.18 to 5.20 demonstrate the effectiveness of this clustering method. See Urquhart (1982) for supplementary reading.

5.6 MIXTURE STATISTICS AND UNSUPERVISED LEARNING

Consider a probability density function that is a mixture of other probability functions. This probability function can then be expressed as

$$p(\underline{x}) = \sum_{i=1}^{M} p(\omega_i)p(\underline{x}|\omega_i) \tag{5.55}$$

$$\sum_{i=1}^{M} p(\omega_i) = 1 \tag{5.56}$$

$$p(\omega_i) \geqslant 0 \quad \forall \; i, \; i = 1, 2, \ldots, M \tag{5.57}$$

where $p(\underline{x}|\omega_i)$ is the likelihood function of ω_i as defined previously and can be interpreted as the probability of $\underline{x}$ given that the state of nature is ω_i. $p(\omega_i)$ denotes the a priori probability of $\underline{x}$ falling in subset S_i, which is later classified as ω_i. $P(\underline{x})$ can then be interpreted as the probability of the unknown pattern samples expressed in terms of the statistics of the natural clusters of those samples. If

$$p(\underline{x}|\omega_i)p(\underline{x}|\omega_j) \simeq 0 \quad \forall \quad \underline{x}, \; j \neq i \qquad (5.58)$$

that is, little overlap (or nearly no overlap) exists between clusters, then Eq. (5.55) becomes approximately

$$p(\underline{x}) = p(\omega_i)p(\underline{x}|\omega_i) \quad \text{if } \underline{x} \in S_i \qquad (5.59)$$

where

$$S_i = \{\underline{x}\,|\,p(\omega_i)p(\underline{x}|\omega_i) \geqslant p(\omega_j)p(\underline{x}|\omega_j) \quad \forall \quad j\}$$

If the overlap is not too great among the component density functions of the mixture, there will exist a one-to-one correspondence between the modes (or local maximums) of the mixture and the individual component class density functions, as shown in Fig. 5.21.

The clustering problem is now to locate the modes (or local maximums) of the probability density $p(\underline{x})$, and further to represent each mode by $p(\underline{x}|\omega_i)$, $i = 1, 2, \ldots$; that is, we are to learn the distribution of $\underline{x}$ without supervision. Locations of these modes in $p(\underline{x})$ might be used for the center-distance definition of clusters.

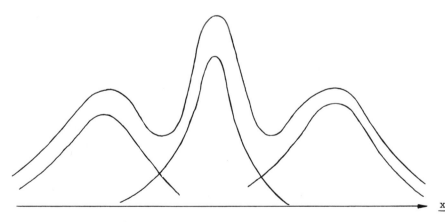

FIG. 5.21 Composite probability density function.

Several procedures can be used in the multimodal search. Random search is one procedure, gradient search is another. But if the statistics of each class and the number of classes are not all known, we must search for some other procedure to use for this problem.

For simplicity, the normal distribution of n dimensions with zero mean is assumed for $p(\underline{x})$, or

$$p(\underline{x}) = \frac{1}{(2\pi)^{n/2}|C|^{1/2}} \exp\left(-\frac{1}{2}\underline{x}^T C^{-1} \underline{x}\right) \tag{5.60}$$

Since C^{-1} is real and symmetric, Eq. (5.60) can be diagonalized by an orthogonal transformation by choosing eigenvectors of C^{-1} as the new basis vector of ζ so that

$$\underline{x} = \zeta \underline{y} \quad \text{and} \quad \underline{x}^T C^{-1} \underline{x} = \underline{y}^T \Lambda \underline{y} \tag{5.61}$$

where

$$\Lambda = \begin{vmatrix} \lambda_1 & & & 0 \\ & \lambda_2 & & \\ & & \ddots & \\ 0 & & & \lambda_n \end{vmatrix}$$

and λ_i, $i = 1, 2, \ldots, n$, are eigenvalues of C^{-1}. The distribution becomes

$$p(\underline{y}) = \frac{1}{(2\pi)^{n/2}|C|^{1/2}} \exp\left(-\frac{1}{2}\underline{y}^T \Lambda \underline{y}\right) \tag{5.62}$$

The probability for a pattern to fall within the characteristic domain D of this density function is then given by

$$p(\underline{y} \in D) = \frac{1}{(2\pi)^{n/2}|C|^{1/2}} \int_D \exp\left(-\frac{1}{2}\underline{y}^T \Lambda \underline{y}\right) d\underline{y} \tag{5.63}$$

where D is the interior of the quadric defined by

$$\underline{y}^T \Lambda \underline{y} = 1 \tag{5.64}$$

whose center is at the origin and whose principal axes have the same directions as the eigenvectors of C^{-1}. Let

$$\underline{z} = \Lambda^{1/2} \underline{y} = \begin{vmatrix} \sqrt{\lambda_1} & & & & & 0 \\ & \ddots & & & & \\ & & \sqrt{\lambda_i} & & & \\ & & & \ddots & & \\ 0 & & & & & \sqrt{\lambda_n} \end{vmatrix} \underline{y}$$

Equation (5.63) can be further reduced to

$$p(\underline{y} \in D) = \frac{\Pi_{i=1}^n \sqrt{\lambda_i}}{(2\pi)^{n/2} |C|^{1/2}} \int_\phi \exp\left(-\frac{1}{2} \underline{z}^T \underline{z}\right) dz = \text{const.} = \alpha$$

(5.65)

where ϕ, the region over which the integral operates, is the circle of unit radius defined by

$$\underline{z}^T \underline{z} = 1 \quad \text{and} \quad |C|^{1/2} = \prod_{i=1}^n \sqrt{\lambda_i}$$

(5.66)

Equation (5.59) becomes

$$p(\underline{x}) = \alpha p(\omega_i)$$

(5.67)

Then we have

$$\frac{p(\omega_1)}{p(\underline{x} \in D_1)} = \frac{p(\omega_2)}{p(\underline{x} \in D_2)} = \cdots = \frac{p(\omega_M)}{p(\underline{x} \in D_M)}$$

(5.68)

and

$$\sum_{i=1}^M p(\omega_i) = 1$$

(5.69)

If N_i, i = 1, 2, ..., represent the number of patterns falling in the modal domain D_i, then the probability $p(\underline{x} \in D_i)$ can be estimated from N_i/N, where N is the number of patterns used for the test and Eq. (5.68) becomes

$$\frac{p(\omega_1)}{N_1} = \frac{p(\omega_2)}{N_2} = \cdots = \frac{p(\omega_M)}{N_M}$$

$$\sum_{i=1}^{M} p(\omega_i) = 1$$

The problem remaining to be solved is how to locate the modes. The geometrical properties of the modal domain with multivariate gaussian density function have been studied. Analysis of the variations of the mean value of the function $p(\underline{x})$ within a suitable domain can be used to determine the local convexity of the function $p(\underline{x})$ at point $\underline{x}$. See Postaire and Vasseur (1981) for more detailed description of this technique.

5.7 CONCLUDING REMARKS

Clustering, which is a very powerful tool in data classification, is an unsupervised approach. Compared to the supervised approach, this approach is less restricted subjectively by prior knowledge. Appropriate application of this natural clustering sometimes produces unexpected inspiration and innovation.

Clustering can be used for training, classification, and mode location as well as for learning. The importance of this approach to classification is considerable. Much has been achieved recently using this line of approach.

It is worthwhile to mention here that the contrast and distinctness of an image can be greatly improved through the use of clustering. More details on this method are given in our discussion of image enhancement by clustering in Chap. 8.

PROBLEMS

5.1 Consider the following samples:

$\underline{x}_1 = (6,0)$ $\underline{x}_5 = (-7,-4)$ $\underline{x}_9 = (-7,6)$

$\underline{x}_2 = (-4,-5)$ $\underline{x}_6 = (7,1)$ $\underline{x}_{10} = (7,0)$

$\underline{x}_3 = (-4,6)$ $\underline{x}_7 = (-5,-4)$ $\underline{x}_{11} = (-5,8)$

$\underline{x}_4 = (-6,6)$ $\underline{x}_8 = (9,1)$ $\underline{x}_{12} = (-5,-6)$

$$\underline{x}_{13} = (-6,9) \qquad \underline{x}_{17} = (9,0) \qquad \underline{x}_{21} = (-6,-6)$$

$$\underline{x}_{14} = (-6,-7) \qquad \underline{x}_{18} = (-7,-5) \qquad \underline{x}_{22} = (-7,8)$$

$$\underline{x}_{15} = (8,-1) \qquad \underline{x}_{19} = (-8,7) \qquad \underline{x}_{23} = (-5,7)$$

$$\underline{x}_{16} = (-7,-6) \qquad \underline{x}_{20} = (7,-1) \qquad \underline{x}_{24} = (8,2)$$

Determine the cluster centers by means of Minimization of Sum of Squared Distances algorithm. Arbitrarily choose $\underline{z}_1(0) = (8,8)$, $\underline{z}_2(0) = (-8,-8)$, and $\underline{z}_3(0) = (-8,8)$.

5.2   Repeat Problem 5.1 using the following pattern samples:

$$\underline{x}_1 = (0,0) \qquad \underline{x}_8 = (2,2) \qquad \underline{x}_{15} = (7,7)$$

$$\underline{x}_2 = (1,0) \qquad \underline{x}_9 = (8,5) \qquad \underline{x}_{16} = (8,7)$$

$$\underline{x}_3 = (0,1) \qquad \underline{x}_{10} = (6,6) \qquad \underline{x}_{17} = (7,8)$$

$$\underline{x}_4 = (1,1) \qquad \underline{x}_{11} = (7,6) \qquad \underline{x}_{18} = (8,8)$$

$$\underline{x}_5 = (2,1) \qquad \underline{x}_{12} = (8,6) \qquad \underline{x}_{19} = (9,8)$$

$$\underline{x}_6 = (3,1) \qquad \underline{x}_{13} = (9,6) \qquad \underline{x}_{20} = (8,9)$$

$$\underline{x}_7 = (0,2) \qquad \underline{x}_{14} = (6,7)$$

Two classes are assumed and the cluster centers can be chosen arbitrarily.

5.3   Repeat Problem 5.2 using the ISODATA algorithm. Start the procedure with one cluster center.

5.4   Use the minimal spanning tree approach to cluster the following data:

$$\underline{x}_1 = (1,6) \qquad \underline{x}_{12} = (10,5) \qquad \underline{x}_{23} = (10,4)$$

$$\underline{x}_2 = (7,6) \qquad \underline{x}_{13} = (11,5) \qquad \underline{x}_{24} = (11,4)$$

$$\underline{x}_3 = (10,7) \qquad \underline{x}_{14} = (12,5) \qquad \underline{x}_{25} = (12,4)$$

$$\underline{x}_4 = (11,6) \qquad \underline{x}_{15} = (13,5) \qquad \underline{x}_{26} = (14,4)$$

$$\underline{x}_5 = (12,6) \qquad \underline{x}_{16} = (14,5) \qquad \underline{x}_{27} = (15,4)$$

$$\underline{x}_6 = (13,6) \qquad \underline{x}_{17} = (15,5) \qquad \underline{x}_{28} = (16,4)$$

$$\underline{x}_7 = (14,6) \qquad \underline{x}_{18} = (2,4) \qquad \underline{x}_{29} = (1,3)$$

$$\underline{x}_8 = (1,5) \qquad \underline{x}_{19} = (3,4) \qquad \underline{x}_{30} = (2,3)$$

$$\underline{x}_9 = (3,5) \qquad \underline{x}_{20} = (5,4) \qquad \underline{x}_{31} = (4,3)$$

$$\underline{x}_{10} = (4,5) \qquad \underline{x}_{21} = (6,4) \qquad \underline{x}_{32} = (6,3)$$

$$\underline{x}_{11} = (6,5) \qquad \underline{x}_{22} = (7,4) \qquad \underline{x}_{33} = (8,3)$$

$\underline{x}_{34} = (9,3)$ $\underline{x}_{40} = (3,2)$ $\underline{x}_{46} = (4,1)$

$\underline{x}_{35} = (11,3)$ $\underline{x}_{41} = (5,2)$ $\underline{x}_{47} = (6,1)$

$\underline{x}_{36} = (12,3)$ $\underline{x}_{42} = (7,2)$ $\underline{x}_{48} = (11,1)$

$\underline{x}_{37} = (13,3)$ $\underline{x}_{43} = (10,2)$ $\underline{x}_{49} = (12,1)$

$\underline{x}_{38} = (14,3)$ $\underline{x}_{44} = (11,2)$ $\underline{x}_{50} = (13,0)$

$\underline{x}_{39} = (0,2)$ $\underline{x}_{45} = (14,2)$

(a) How many clusters result?
(b) List the connections that define the main diameter.

part **II**
PREPROCESSING OF DATA FOR PATTERN RECOGNITION

6
Dimensionality Reduction and Feature Selection

6.1 OPTIMAL NUMBER OF FEATURES IN CLASSIFICATION OF MULTIVARIATE GAUSSIAN DATA

The pattern space is usually of high dimensionality. The objective of the feature selection is to reduce the dimensionality of the measurement space to a space suitable for the application of pattern classification algorithms. During this process of feature selection, only the salient features necessary for the recognition process are retained so that classification can be implemented on a vastly reduced feature space. Much work has been done in finding the dependences of the probability of misclassification on the dimensionality of the feature vector, the number of training samples, and the true parameters of the class-conditional densities.

As discussed in Chap. 4, for a two-class problem with $p(\omega_1) = p(\omega_2)$, $\underline{C}_1 = \underline{C}_2 = \underline{C}$, and all parameters, including $\underline{m}_1$ and $\underline{m}_2$, being known, the minimum classification error occurs when a pattern sample $\underline{x}$ is classified such that

$$\underline{x} \in \omega_1 \quad \text{when } \underline{x}^T \underline{C}^{-1}(\underline{m}_1 - \underline{m}_2) - \frac{1}{2}(\underline{m}_1 + \underline{m}_2)^T \underline{C}^{-1}(\underline{m}_1 - \underline{m}_2)$$
$$> 0 \qquad (6.1)$$

$$\underline{x} \in \omega_2 \quad \text{otherwise}$$

But in the case when the parameters are not known, the mean vectors $\underline{m}_1$ and $\underline{m}_2$ and the estimate of the covariance matrix have to be computed from the training samples N_i from class ω_i, $i = 1, 2$.

Unfortunately, evaluation of these quantities is not easy. Various approximations to this statistic have been developed, but they are still very complex in their mathematical forms. Nevertheless, the dependences of the average probability of misclassification on the Mahalanobis distance r^2, the number of samples per class N, and the number of features p are obtained. In general, for a given p, an increase in the values of r^2 and/or N decreases the average error rate. Jain and Walter (1978) have derived an expression for the minimal increase in the Mahalanobis distance needed to keep the misclassification rate unchanged when a new feature is added to the original set of p features:

$$r_{p+1}^2 - r_p^2 = \delta r_p^2 = \frac{r_p^2}{2N - 3 - p} \tag{6.2}$$

where r_p^2 and r_{p+1}^2 represent, respectively, the Mahalanobis distances produced by the p and p + 1 features. Equation (6.2) shows that the minimal increase is a fraction of r_p^2 and that this fraction increases with p and decreases with N, the sample size. The problem of determining the optimal number of features for a given sample size now becomes to find when the contribution of an additional feature to the accumulated Mahalanobis distance is below a threshold. Let the contribution of the ith feature to the Mahalanobis distance be d_i^2; then we have

$$r_p^2 = \sum_{i=1}^{p} d_i^2 \tag{6.3}$$

Assume that the features do not have the same power of discrimination and that the contribution of each feature is a fixed fraction of that of the previous feature, such that

$$d_i = \xi d_{i-1} \quad i = 2, \ldots, p \tag{6.4}$$

where ξ is a positive arbitrary constant and less than 1. Substitution of Eq. (6.4) in Eq. (6.3) yields

$$r_p^2 = d_1^2 + \xi^2 d_1^2 + \cdots + \xi^{2p} d_1^2$$

$$= d_1^2 \frac{1 - \xi^{2p}}{1 - \xi^2} \tag{6.5}$$

where d_1^2 is the Mahalanobis distance computed with only the first feature. Since $\xi < 1$, it implies that the features are arranged in a best-to-worst order. We can then determine the smallest set of features by comparing it with the threshold to maximize the classifier's performance.

6.2 FEATURE ORDERING BY MEANS OF CLUSTERING TRANSFORMATION

As discussed in Sec. 6.1, features chosen for classification are usually not of the same significance. Decreasing weights assigned to measurements with decreasing significance can be realized through a linear transformation (Tou and Gonzalez 1974). Let the transformation matrix used for this purpose be a diagonal matrix, or

$$\underline{w} = (w_{jk}) \qquad w_{jk} = \begin{cases} 0 & \text{when } j \neq k \\ \\ w_{jj} & \text{when } j = k \end{cases} \qquad (6.6)$$

where w_{jj}, $j = 1, \ldots, n$, represents the feature-weighting coefficients. Our problem now is to determine the coefficients w_{jj} so that a good clustering can be obtained. Under such circumstances, the intraset distance between pattern points in a set is minimized. The intraset distance $\overline{D}^2$ for pattern points after transformation has already been derived, as shown by Eq. (5.19), which is repeated here with the weighting coefficients added:

$$\overline{D}^2 = 2 \sum_{j=1}^{n} (w_{jj}\sigma_j)^2 \qquad (6.7)$$

where σ_j^2 is the sample variance of the components along the x_j coordinate direction. The Lagrange multiplier can be used for minimization of the intraset distance. Two different constraints can be considered.

Constraint 1 When the constraint is $\sum_{j=1}^{n} w_{jj} = 1$, the minimization of $\overline{D}^2$ is equivalent to the minimization of

$$S = 2 \sum_{j=1}^{n} (w_{jj}\sigma_j)^2 - \rho_1 \left(\sum_{j=1}^{n} w_{jj} - 1 \right) \qquad (6.8)$$

Take the partial derivative of S with respect to w_{jj} and equating it to zero, we have

$$w_{jj} = \frac{\rho_1}{4\sigma_j^2} \tag{6.9}$$

Similarly, taking the partial derivative of S with respect to the Lagrange multiplier ρ_1 and equating it to zero yields

$$\sum_{j=1}^{n} w_{jj} = 1 \tag{6.10}$$

Combining Eqs. (6.9) and (6.10) yields

$$\sum_{j=1}^{n} \frac{\rho_1}{4\sigma_j^2} = 1 \tag{6.11}$$

or

$$\rho_1 = \frac{4}{\sum_{j=1}^{n} \sigma_j^{-2}} \tag{6.12}$$

Substitution of Eq. (6.12) back into Eq. (6.9) gives the feature weighting coefficients

$$w_{jj} = \frac{1}{\sigma_j^2 \sum_{j=1}^{n} \sigma_j^{-2}} \tag{6.13}$$

From Eq. (6.13) it can be seen that the value under the summation sign in the denominator is the same for all w_{jj}, $j = 1, \ldots, n$, and therefore w_{jj} varies inversely with σ_j^2.

Constraint 2 When the constraint is $\prod_{j=1}^{n} w_{jj} = 1$, the minimization of $\overline{D}^2$ is equivalent to the minimization of

$$S = 2 \sum_{j=1}^{n} (w_{jj}\sigma_j)^2 - \rho_2 \left(\prod_{j=1}^{n} w_{jj} - 1 \right) \tag{6.14}$$

Taking the partial derivative of S with respect to w_{jj} and equating it to zero yields

$$4w_{jj}\sigma_j^2 - \rho_2 \prod_{\substack{k=1 \\ k \neq j}}^{n} w_{kk} = 0 \tag{6.15}$$

Multiplying both sides by w_{jj}, we have

$$4w_{jj}^2\sigma_j^2 - \rho_2 \prod_{k=1}^{n} w_{kk} = 0 \tag{6.16}$$

Substitution of $\prod_{j=1}^{n} w_{jj} = 1$ into Eq. (6.16) gives

$$4w_{jj}^2\sigma_j^2 - \rho_2 = 0 \tag{6.17}$$

or

$$w_{jj} = \frac{\sqrt{\rho_2}}{2\sigma_j} \tag{6.18}$$

Similarly,

$$\frac{\partial S}{\partial \rho_2} = \prod_{j=1}^{n} w_{jj} - 1 = 0 \tag{6.19a}$$

or

$$\prod_{j=1}^{n} w_{jj} = 1 \tag{6.19b}$$

which satisfies the given constraint. Substituting Eq. (6.18) into Eq. (6.9b) yields

$$\prod_{j=1}^{n} \frac{\sqrt{\rho_2}}{2\sigma_j} = 1 \tag{6.20}$$

or

$$\frac{(\rho_2)^{n/2}}{2^n \prod_{j=1}^{n} \sigma_j} = 1 \tag{6.21}$$

After simplification, we obtain

$$\rho_2 = 4\left(\prod_{j=1}^{n} \sigma_j\right)^{2/n} \tag{6.22}$$

Combining Eqs. (6.18) and (6.22) yields

$$w_{jj} = \frac{1}{\sigma_j}\left(\prod_{j=1}^{n} \sigma_j\right)^{1/n} \tag{6.23}$$

Note that the continual product inside the parentheses is the same for all w_{jj}, $j = 1, \ldots, n$, and therefore w_{jj} varies inversely with σ_j.

Although the results obtained are somewhat different for different constraints on w_{jj}, the guides for choosing the feature-weighting coefficients are the same for both cases. That is, a small weight is to be assigned to a feature of large variation, whereas a feature with a small standard deviation σ_j will be weighted heavily. A feature with a small standard deviation σ_j implies that it is more reliable. It is desirable that the more reliable features be more heavily weighted.

6.3 CANONICAL ANALYSIS AND ITS APPLICATIONS TO REMOTE SENSING PROBLEMS

The objective of canonical analysis (McMurty 1976) is to derive a linear transformation that will emphasize the differences among the pattern samples belonging to different categories. In other words, the objective is to define new coordinate axes in directions of high information content useful for classification purposes.

For multispectral remote sensing applications, each observation will be represented as an n-component vector,

$$\underline{x}_{ij} = \begin{vmatrix} x_{ij1} \\ \cdot \\ \cdot \\ \cdot \\ x_{ijk} \\ \cdot \\ \cdot \\ \cdot \\ x_{ijn} \end{vmatrix} \tag{6.24}$$

where n represents the dimensionality of the observation vector $\underline{x}_{ij}$, or the number of channels used for the observation. x_{ijk} represents the observation (or the intensity of picture element) in the kth channel for picture element j in scan line i. Let $\hat{\underline{m}}_\ell$ and $\hat{\underline{C}}_\ell$ denote, respectively, the sample mean vector and the covariance matrix for the ℓth category (ℓ = 1, 2, ..., M). These two quantities can be obtained from the training set.

Our problem now is to find a transformation matrix. By means of this transformation, two results are expected. First, the n-dimensional observation vector $\underline{x}$ will be transformed into a new vector $\underline{y}$ with a dimensionality p which is less than n; or

$$\underline{y}_{ij} = \underline{C}\underline{x}_{ij} \tag{6.25}$$

where $\underline{C}$ is a p × n transformation matrix and $\underline{y}_{ij}$ will be represented as

$$\underline{y}_{ij} = \begin{vmatrix} y_{ij1} \\ \cdot \\ \cdot \\ \cdot \\ y_{ijk} \\ \cdot \\ \cdot \\ \cdot \\ y_{ijp} \end{vmatrix} \tag{6.26}$$

Second, the transformation matrix $\underline{C}$ is chosen such that $\underline{C}\underline{A}\underline{C}^T$ will be a diagonal matrix whose diagonal elements are the variances of the transformed variables and are arranged in descending order. The transformed variables with the largest values can be assumed to have the greatest discriminatory power, since they show the greatest spread (variance) among categories.

The matrix A mentioned above is the covariance of the means of the categories (referred to as the "among"-categories covariance matrix) and represents the spread in n-dimensional space among the categories. It can be defined as

$$\underline{A} = \underline{X}\underline{N}\underline{X}^T - \frac{1}{\sum_{\ell=1}^{M} n_\ell} (\underline{X}\underline{n})(\underline{X}\underline{n})^T \tag{6.27}$$

where n_ℓ is the number of observations for category ℓ; M is the number of categories, and $\underline{X}$ is defined as an n × M matrix of all the category means composed of all mean vectors $\hat{\underline{m}}_k$, k = 1, 2, ..., M. as

$$\underline{X} = [\hat{\underline{m}}_1, \hat{\underline{m}}_2, \ldots, \hat{\underline{m}}_M] = \begin{array}{c} \text{ch. 1} \\ \text{ch. 2} \\ \vdots \\ \text{ch. n} \end{array} \begin{array}{c} \overset{\text{cat. 1}}{\begin{vmatrix} m_{11} \\ m_{21} \\ \vdots \\ m_{n1} \end{vmatrix}} \end{array} \begin{array}{c} \overset{\text{cat. 2}}{} \\ m_{12} \\ m_{22} \\ \vdots \\ m_{n2} \end{array} \begin{array}{c} \overset{\cdots \text{ cat. M}}{} \\ \cdots \; m_{1M} \\ \cdots \; m_{2M} \\ \; \vdots \\ \cdots \; m_{nM} \end{array} \qquad (6.28)$$

The $\underline{N}$ and $\underline{n}$ in Eq. (6.27) are, respectively, an $M \times M$ matrix and an $M \times 1$ vector of the number of observations in the categories as

$$\underline{N} = \begin{vmatrix} n_1 & & & & 0 \\ & n_2 & & & \\ & & \cdot & & \\ & & & \cdot & \\ 0 & & & & n_M \end{vmatrix} \qquad (6.29)$$

and

$$\underline{n} = \begin{vmatrix} n_1 \\ n_2 \\ \cdot \\ \cdot \\ \cdot \\ n_M \end{vmatrix} \qquad (6.30)$$

Let W be the combined covariance matrix for all the categories (referred to as the "within"-categories covariance matrix), which can be computed as

$$W = \frac{\sum_{\ell=1}^{M} (n_\ell - 1)\hat{C}_\ell}{\sum_{\ell=1}^{M} n_\ell - M} \qquad (6.31)$$

where $\hat{C}_\ell$ is the covariance matrix for category ℓ, n_ℓ is the number of observations for category ℓ, and M is the number of categories.

The matrix C can be made to be unique only if the following constraint is placed on it:

$$CWC^T = I \tag{6.32}$$

where I is a p × p identity matrix. Let a new matrix $W^{1/2}$ be such defined that

$$W^{1/2}(W^{1/2})^T = W \tag{6.33}$$

and

$$W^{-1/2} = (W^{1/2})^{-1} \tag{6.34}$$

where $(W^{1/2})^{-1}$ is the inverse of $W^{1/2}$. Then

$$
\begin{aligned}
CAC^T &= CW^{1/2}W^{-1/2}A(W^{-1/2})^T(W^{1/2})^T C^T \\
&= (CW^{1/2})W^{-1/2}A(W^{-1/2})^T(CW^{1/2})^T
\end{aligned} \tag{6.35}
$$

and Eq. (6.32) becomes

$$CWC^T = I = (CW^{1/2})(W^{1/2})^T C^T = (CW^{1/2})(CW^{1/2})^T$$

or

$$FF^T = I \tag{6.36}$$

if $CW^{1/2}$ is replaced by F. Equation (6.35) then becomes

$$CAC^T = FVF^T = \Lambda \tag{6.37}$$

where V is used to substitute for $W^{-1/2}A(W^{-1/2})^T$ in Eq. (6.35). Our problem now becomes one of finding F to diagonalize matrix V subject to the constraint $FF^T = I$. Before we can do this, we must first find $W^{1/2}$. Let us construct matrices D and E such that

$$W = EDE^T \tag{6.38}$$

and

$$EE^T = I \tag{6.39}$$

where D is a diagonal matrix whose diagonal elements are the eigen-values of W, and E is the matrix whose columns are the normalized eigen-vectors of W. Then we have

$$W^{1/2} = ED^{1/2}E^T$$

and   (6.40)

$$W^{-1/2} = ED^{-1/2}E^T$$

where $D^{1/2}$ is defined as a diagonal matrix whose diagonal elements are the square roots of the corresponding elements in D, and $D^{-1/2}$ is similarly defined.

Once $W^{-1/2}$ has been computed, $V = W^{-1/2}A(W^{-1/2})^T$ may be de-termined. Then the problem becomes one of finding Λ and F from Eq. (6.37) subject to the constraint of Eq. (6.36). That is, Λ is the di-agonal matrix of eigenvalues of V, and F is the matrix whose rows are the corresponding normalized eigenvectors. These matrices are then as follows:

$$\Lambda = \begin{bmatrix} \lambda_1 & & & & & 0 & \\ & \lambda_2 & & & & & 0 \\ 0 & & \ddots & & & & \\ & & & \lambda_p & & & \\ \hline & & & & \ddots & & \\ 0 & & & & & \lambda_n \end{bmatrix} \qquad (6.41a)$$

and

$$F = [f_1, f_2, \ldots, f_n] \qquad (6.41b)$$

Only the p × p submatrix Λ^* of Λ contains the distinguishable eigen-values such that

$$\Lambda^* = \begin{bmatrix} \lambda_1 & & 0 \\ & \ddots & \\ 0 & & \lambda_p \end{bmatrix} \qquad (6.42)$$

with $\lambda_{p+1} = \cdots = \lambda_n = 0$. This will be used as a discriminant space. In a similar manner, F is partitioned as

$$F^* = [f_1, f_2, \ldots, f_p] \tag{6.43}$$

As a result of this partitioning, the transformation matrix C becomes C*, such that

$$C^* = F^*W^{-1/2} \tag{6.44}$$

which is now a p × n matrix.

6.4 NONPARAMETRIC FEATURE SELECTION METHOD APPLICABLE TO MIXED FEATURES

The feature selection methods discussed previously are for those features that are usual quantitative variables. In this section we discuss a nonparametric feature selection method that can be applied to pattern recognition problems based on mixed features. That is, some of the features are quantitative, whereas others are qualitative.

Feature selection for this purpose based on local interclass structure was suggested by Ichino (1981). This method consists of the following three steps:

1. Divide the pattern space into a set of subregions, or, in basic event generation (BEG) terminology, generate the basic events E_{ik}, $k = 1, \ldots, N_{ei}$, for class ω_i by means of the BEG algorithm.
2. Use the set theoretic feature selection method to find a subset of features, F_{ik}, for each subregion (i.e., for each basic event E_{ik}).
3. Construct the desired feature subset by taking the union of feature subsets obtained in step 2 as $\cup_{k=1}^{N_{ei}} F_{ik}$.

The basic event generation algorithm used in step 1 is essentially a merging process. Let $\underline{x} = (x_1, x_2, \ldots, x_n)$ be a pattern vector in the pattern space $\underline{R}^n = R \times R \times \cdots \times R$. Then if $\underline{E}_1$ and $\underline{E}_2$ are two events in R^n, it is obvious that "merge" of these two events is also in R^n; thus

$$M(\underline{E}_1, \underline{E}_2) = E \subset R^n \tag{6.45}$$

where $M(\cdot, \cdot)$ represents the merging function. Suppose that we have training samples $\underline{x}_1, \underline{x}_2, \ldots, \underline{x}_{N_i}$ from class ω_i, and training samples

$\underline{y}_1, \underline{y}_2, \ldots, \underline{y}_{N_j}$ from classes other than class ω_i. The events generated by the BEG algorithm for class ω_i will be $\underline{E}_{ik}$, $k = 1, 2, \ldots,$ N_{e_i} ($N_{e_i} \leqslant N_i$). Then we have

$$\underline{x}_\ell \in \bigcup_{k=1}^{N_{e_i}} \underline{E}_{ik} \qquad \ell = 1, 2, \ldots, N_i \qquad (6.46)$$

and

$$\text{dist}(\underline{y}_\ell | \underline{E}_{ik}) \geqslant T_A \qquad \ell = 1, 2, \ldots, N_j; \quad k = 1, 2, \ldots, N_{e_i} \qquad (6.47)$$

where $\text{dist}(\underline{y}_\ell | \underline{E}_{ik})$, the distance between $\underline{y}_\ell$ and $\underline{E}_{ik}$, must be greater than or equal to T_A, a certain positive number that is usually chosen to be 1. If $\underline{E}_{ik}$ and $\underline{y}_\ell$ are expressed, respectively, as

$$\underline{E}_{ik} = E_{ik}^1 \times E_{ik}^2 \times \cdots \times E_{ik}^n \qquad (6.48)$$

and

$$\underline{y}_\ell = (y_{\ell 1}, y_{\ell 2}, \ldots, y_{\ell n}) \qquad (6.49)$$

then $\text{dist}(\underline{y}_\ell | \underline{E}_{ik})$ can be defined as

$$\text{dist}(\underline{y}_\ell | \underline{E}_{ik}) = \sum_{p=1}^{n} \phi(y_{\ell p} | E_{ik}^p) \qquad (6.50)$$

where

$$\phi(y_{\ell p} | E_{ik}^p) = \begin{cases} 1 & \text{if } y_{\ell p} \in / \ E_{ik}^p \\ \\ 0 & \text{otherwise} \end{cases} \qquad (6.51)$$

After the basic events $\underline{E}_{ik}$, $k = 1, 2, \ldots, N_{e_i}$, are generated by the BEG algorithm for class ω_i, the next step will be to select a minimum number of features F_{ik}, by which the basic events $\underline{E}_{ik}$ for class ω_i can be separated from the training samples drawn from classes other than ω_i. F_{ik} is said to be the minimum feature subset for the event $\underline{E}_{ik}$ if the following two conditions are satisfied:

(1) $\text{dist}(\underline{y}_\ell | \underline{E}_{ik})_{F_{ik}} \geqslant T_A$ $\ell = 1, 2, \ldots, N_j$ (6.52)

(2) $\text{dist}(\underline{y}_\ell | \underline{E}_{ik})_{F_{ik}-[p]} < T_A$ for some values of ℓ (6.53)

where p is a feature in F_{ik}. Equations (6.52) and (6.53) are evaluated respectively with feature subsets F_{ik} and $F_{ik} - [p]$. Conditions 1 and 2 are equivalent, respectively, to

$$|F_{ik} \cap F_{ik}^\ell| \geqslant T_A \qquad \ell = 1, 2, \ldots, N_j \qquad (6.54)$$

and

$$|(F_{ik} - [p]) \cap F_{ik}^\ell| < T_A \qquad \text{for some values of } \ell \qquad (6.55)$$

where the F_{ik}^ℓ represent the sets of effective features for ℓ training samples not in class ω_j. A sequential search procedure for this feature selection can be summarized as follows:

1.   Let F_{ik} be the original feature set F_0.
2.   For a feature $u \in F_{ik}$, if $|(F_{ik} - [u]) \cap F_{ik}^\ell| \geqslant T_A$, $\ell = 1, 2, \ldots, N_j$, discard the feature and replace F_{ik} by $F_{ik} - [u]$; otherwise, try for other features in F_{ik}.
3.   Iterate step 2 until no feature in F_{ik} can be removed.

The resultant F_{ik} obtained by this procedure is a miniumu feature subset. For example, the original feature set is chosen to be $F_0 = [a,b,c,d]$. Let $F_{ik}^1 = [b,c]$, $F_{ik}^2 = [a]$, $F_{ik}^3 = [a]$, and $F_{ik}^4 = [b,d]$ be the sets of effective features for the given four training samples $\underline{y}_1$, $\underline{y}_2$, $\underline{y}_3$, and $\underline{y}_4$, respectively. According to the sequential search procedure, either features c and d or feature b can be discarded, but not both. The minimum feature subset is then $[a,b]$ or $[a,c,d]$.

7
Image Transformation

The necessity of performing pattern recognition problems by computer lies on the large set of data to be dealt with. The preprocessing of large volumes of these data into a better form will be very helpful for more accurate pattern recognition. A two-dimensional image is a very good example of problems with a large data set. The preprocessing of an image can be carried out in one of two domains: the spatial domain and the transform domain. Figure 7.1 shows the general configuration of digital image processing in the spatial domain or in the Fourier domain and the relationship between them.

When an image is processed in the spatial domain, the processing of digitized image is carried out directly either by point processing or by neighborhood processing for enhancement or restoration. But if an image is to be processed in the transform domain, the digitized image will first be transformed by discrete Fourier transform (DFT) or fast Fourier transform (IFFT)] will be carried out on the result to trans- processed, an inverse operation of the FFT [called the inverse fast fourier transform (IFFT)] will be carried out on the result to transform it back to an image in the spatial domain.

Image transform and image inverse transform are two intermediate processes. They are linked such that image processing can be carried out in the transform domain instead of in the spatial domain. In so doing, three objectives are expected:

1. Processing might be facilitated in the transform domain for some operations, such as convolution and correlation.
2. Some features might be more obvious and easier to extract in that domain.

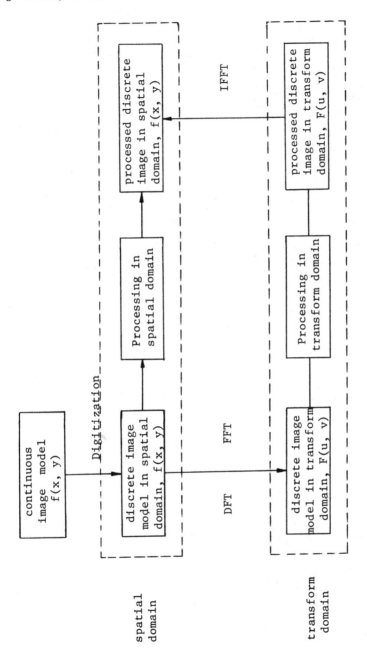

FIG. 7.1 Schematic diagram of image processing.

3. Data compression might be possible, thus reducing the on-line and off-line storage requirements and also the bandwidth requirements in transmission.

7.1 FORMULATION OF THE IMAGE TRANSFORM

If a continuous function $f(x)$ as shown in Fig. 7.2 is discretized into N samples Δx apart, such as $f(x_0)$, $f(x_0 + \Delta x)$, $f(x_0 + 2\Delta x)$, ..., $f(x_0 + (N - 1)\Delta x)$, the function $f(x)$ can be expressed as

$$f(x) = f(x_0 + x\,\Delta x) \quad x = 0, 1, \ldots, N - 1 \tag{7.1}$$

With this in mind we have the discrete Fourier transform pair*

$$F(u) = \frac{1}{N} \sum_{x=0}^{N-1} f(x)e^{-j2\pi ux/N} \quad u = 0, 1, \ldots, N - 1 \tag{7.2}$$

$$f(x) = \sum_{u=0}^{N-1} F(u)e^{j2\pi ux/N} \quad x = 0, 1, \ldots, N - 1 \tag{7.3}$$

where N is the number of samples taken from the function curve. This corresponds to the transform pair for a continuous one-dimensional function:

$$F(j\omega) = \int_{-\infty}^{\infty} e^{-j\omega t} f(t)\, dt \tag{7.4}$$

$$f(t) = \frac{1}{2\pi} \int_{-\infty}^{\infty} e^{j\omega t} F(j\omega)\, d\omega \tag{7.5}$$

Extending this to two-dimensional functions, we have the Fourier transform pair for a continuous function as

$$F(u,v) = \int_{-\infty}^{\infty}\!\!\int f(x,y)e^{-j2\pi(ux+vy)}\, dx\, dy \tag{7.6}$$

and

---

*A proof of these results is very lengthy and is beyond the scope of this book. Details of the derivation can be found in Brigham (1974).

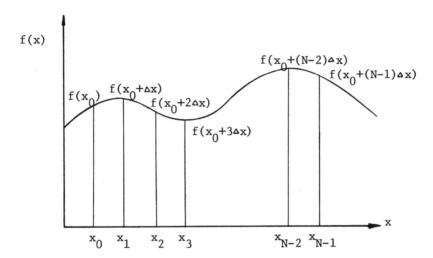

FIG. 7.2 Discretization of a continuous function.

$$f(x,y) = \int_{-\infty}^{\infty}\int F(u,v)e^{j2\pi(ux+vy)} \, du \, dv \qquad (7.7)$$

where x and y are spatial coordinates and $f(x,y)$ is the image model; while u and v are the spatial frequencies and $F(u,v)$ is the frequency spectrum. The corresponding discrete transform pair for the two-dimensional function will be

$$F(u,v) = \frac{1}{N^2} \sum_{x=0}^{N-1}\sum_{y=0}^{N-1} f(x,y)e^{-j2\pi(ux+vy)/N}$$

$$u, v = 0, 1, 2, \ldots, N-1 \qquad (7.8)$$

and

$$f(x,y) = \sum_{u=0}^{N-1}\sum_{v=0}^{N-1} F(u,v)e^{j2\pi(ux+vy)/N}$$

$$x, y = 0, 1, 2, \ldots, N-1 \qquad (7.9)$$

The frequency spectrum $F(u,v)$ can be computed if the appropriate values of x, y, u, v, and $f(x,y)$ are substituted into Eq. (7.8). Clearly, the computation is rather cumbersome, and spectrum computation by computer is suggested when N becomes large.

There are quite a few transformation techniques available, among them the Fourier transform represented by Eq. (7.8). If the exponential factor $e^{-j2\pi(ux+vy)/N}$ is replaced by a more general function, $g(x,y;u,v)$, Eq. (7.8) becomes

$$F(u,v) = \sum_{x=0}^{N-1} \sum_{y=0}^{N-1} f(x,y)g(x,y;u,v) \qquad (7.10)$$

where $F(u,v)$ is an $N \times N$ transformed image array if $f(x,y)$ is an $N \times N$ array of numbers used to represent the discrete image model as follows:

$$[f] = \begin{bmatrix} f(0,0) & f(0,1) & \cdots & f(0, N-1) \\ \vdots & & & \\ f(N-1) & \cdots & & f(N-1, N-1) \end{bmatrix} \qquad (7.11)$$

The function $g(x,y;u,v)$ in Eq. (7.10) is the forward transform kernel. Correspondingly, we can write its inverse transform:

$$f(x,y) = \sum_{u=0}^{N-1} \sum_{v=0}^{N-1} F(u,v)h(x,y;u,v) \qquad (7.12)$$

where $h(x,y;u,v)$ is the inverse transform kernel. For the case of Fourier transform, the inverse transform kernel is $e^{j2\pi(ux+vy)}$. Equations (7.10) and (7.12) form a transform pair. The transformation is unitary* if the following orthonormality conditions are met:

$$\sum_{u} \sum_{v} g(x,y;u,v)g^*(x_0,y_0;u,v) = \delta(x - x_0, y - y_0) \qquad (7.13)$$

$$\sum_{u} \sum_{v} h(x,y;u,v)h^*(x_0,y_0;u,v) = \delta(x - x_0, y - y_0) \qquad (7.14)$$

$$\sum_{x} \sum_{y} g(x,y;u,v)g^*(x,y;u_0,v_0) = \delta(u - u_0, v - v_0) \qquad (7.15)$$

---

*A is said to be a *unitary matrix* if the matrix inverse is given by $A^{-1} = A^{*T}$. A real unitary matrix is called an *orthogonal matrix*. For such a matrix, $A^{-1} = A^T$.

$$\sum_x \sum_y h(x,y;u,v)h^*(x,y;u_0,v_0) = \delta(u - u_0, v - v_0) \qquad (7.16)$$

where the superscript * denotes a complex conjugate and the Dirac delta function is

$$\delta = \begin{cases} \infty & \text{at } x = x_0, \ y = y_0 \\ \\ 0 & \text{elsewhere} \end{cases} \qquad (7.17)$$

or

$$\delta = \begin{cases} \infty & \text{at } u = u_0, \ v = v_0 \\ \\ 0 & \text{elsewhere} \end{cases} \qquad (7.18)$$

These can easily be proved by substituting $e^{-j2\pi(ux+vy)/N}$ and $e^{+j2\pi(ux_0+vy_0)/N}$, respectively for $g(x,y;u,v)$ and $g^*(x_0,y_0;u,v)$ in Eq. (7.13).

Two-dimensional transformation is a very tedious mathematical operation, and therefore lots of effort have been spent in simplifying it. The separability property of the transformation is very effective for this purpose. The transformation is "separable" if its kernel can be written as

$$g(x,y;u,v) = g_{col}(x,u)g_{row}(y,v) \qquad \text{forward transform} \qquad (7.19)$$

$$h(x,y;u,v) = h_{col}(x,u)h_{row}(y,v) \qquad \text{inverse transform} \qquad (7.20)$$

A separable unitary transform can thus be computed in two steps:

1. Transform column-wise or one-dimensional transform along each column of the image $f(x,y)$:

$$P(u,y) = \sum_{x=0}^{N-1} f(x,y)g_{col}(x,u) \qquad (7.21)$$

where $g_{col}(x,u)$ is the forward column transform kernel and is $e^{-j2\pi ux/N}$ for the Fourier transform

2. Transform row-wise, or one-dimensional unitary transform along each row of $P(u,y)$:

$$F(u,v) = \sum_{y=0}^{N-1} P(u,y)g_{row}(y,v) \qquad (7.22)$$

where $g_{row}(y,v)$ is the forward row transform kernel and is $e^{-j2\pi vy/N}$ for the Fourier transform. Thus a two-dimensional transform may be computed in two steps, each being a one-dimensional transform. If an efficient and effective one-dimensional transform algorithm is set up, it can be used repeatedly for a two-dimensional transformation.

7.2 FUNCTIONAL PROPERTIES OF THE TWO-DIMENSIONAL FOURIER TRANSFORM

As mentioned in Sec. 7.1, the Fourier transform for a continuous function is

$$F(u,v) = \int_{-\infty}^{\infty}\int f(x,y)e^{-j2\pi(ux+vy)} \, dx \, dy \qquad (7.23)$$

This transform function is in general complex and consists of two parts, such as

$$F(u,v) = \text{Real}(u,v) + j\,\text{Imag}(u,v) \qquad (7.24)$$

or in polar form,

$$F(u,v) = |F(u,v)|\phi(u,v) \qquad (7.25)$$

where $F(u,v) = [\text{Real}^2(u,v) + \text{Imag}^2(u,v)]^{1/2}$ = Fourier spectrum of $f(x,y)$,

$$\phi(u,v) = \tan^{-1}\left[\frac{\text{Imag}(u,v)}{\text{Real}(u,v)}\right] = \text{phase angle}$$

and $|F(u,v)|^2 = E(u,v)$ = energy spectrum of $f(x,y)$.

The inverse transform of $F(u,v)$ gives $f(x,y)$. The inverse transform forms a pair with Eq. (7.23).

$$f(x,y) = \mathcal{F}^{-1}[F(u,v)] = \int_{-\infty}^{\infty}\int F(u,v)e^{j2\pi(ux+vy)} \, du \, dv \qquad (7.26)$$

This transform pair can be shown to exist if $f(x,y)$ is continuous and integrable and $F(u,v)$ is also integrable. If we have a simple rectangular bar object with univorm intensity, that is, $f(x,y) = f_0$, shown shaded on Fig. 7.3, its Fourier spectrum can then be computed according to Eq. (7.23) as follows:

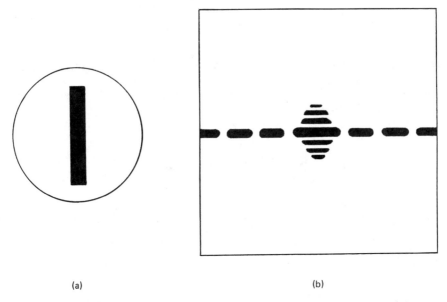

(a)   (b)

FIG. 7.3 Fourier spectrum of a simple rectangular bar object with
uniform intensity: (a) object; (b) spectrum.

$$F(u,v) = \int_{-\infty}^{\infty}\int f(x,y)e^{-j2\pi(ux+vy)}\,dx\,dy$$

$$= f_0\int_0^{x_0} e^{-j2\pi ux}\,dx\int_0^{y_0} e^{-j2\pi vy}\,dy$$

$$= f_0\frac{(e^{-j\pi ux_0} - e^{j\pi ux_0})e^{-j\pi ux_0}}{-j2\pi u}$$

$$\times\frac{(e^{-j\pi vy_0} - e^{j\pi vy_0})e^{-j\pi v_0 y_0}}{-j2\pi v}$$

$$= f_0 x_0 y_0 \frac{\sin\pi ux_0}{\pi ux_0}e^{-j\pi ux_0}\frac{\sin\pi vy_0}{\pi vy_0}e^{-j\pi vy_0} \qquad (7.27)$$

or

$$F(u,v) = f_0 x_0 y_0 \; \mathrm{sinc}(ux_0) e^{-j\pi ux_0} \; \mathrm{sinc}(vy_0) \; e^{-j\pi vy_0} \qquad (7.28)$$

if $\mathrm{sinc}(ux_0)$ and $\mathrm{sinc}(vy_0)$ substitute respectively for $(\sin \pi ux_0)/\pi ux_0$ and $(\sin \pi vy_0)/\pi vy_0$. Figure 7.3b shows the plot of the intensity of the spectrum $F(u,v)$, from which we can see clearly that the spectrum in the intensity plot varies as a sinc function.

Following are some properties of the Fourier transform that are worthy of discussion.

Kernel separability. The Fourier transform given in Eq. (7.23) can be expressed in separate form as follows:

$$F(u,v) = \int_{-\infty}^{\infty} \left[\int_{-\infty}^{\infty} f(x,y) e^{-j2\pi ux} \, dx \right] e^{-j2\pi vy} \, dy \qquad (7.29)$$

The integral in brackets is $F_y(u,y)$ (row-wise transform). $F(u,v)$ can then be expressed as

$$F(u,v) = \int_{-\infty}^{\infty} F_y(u,y) e^{-j2\pi vy} \, dy \qquad (7.30)$$

or

$$F(u,v) = \int_{-\infty}^{\infty} F_x(x,v) e^{-j2\pi ux} \, dx \qquad (7.31)$$

where

$$F_x(x,v) = \int_{-\infty}^{\infty} f(x,y) e^{-j2\pi vy} \, dy \qquad \text{(column-wise transform)}$$

The principal significance of the kernel separability is that a two-dimensional Fourier transform can be separated into two computational steps, each a one-dimensional Fourier transform—which is much less complicated than the two-dimensional transform. Thus

$$\mathcal{F}(u,v) = \mathcal{F}_x \{ \mathcal{F}_y [f(x,y)] \} \qquad (7.32)$$

or

$$F(u,v) = \mathcal{F}_y \{ \mathcal{F}_x [f(x,y)] \} \qquad (7.33)$$

where $\mathcal{F}_x$ and $\mathcal{F}_y$ represent respectively the column-wise and row-wise transformations.

This is also true for the inverse Fourier transform. Since the same one-dimensional Fourier transform is employed in these two steps, more effort can be concentrated on the design of the algorithm to make it more effective.

Linearity. The fourier transform is a linear operator and possesses distributivity and scaling properties. Thus

$$\mathcal{F}[a_1 f_1(x,y) + a_2 f_2(x,y)] = a_1 \mathcal{F}[f_1(x,y)] + a_2 \mathcal{F}[f_2(x,y)]$$

$$= a_1 F_1(u,v) + a_2 F_2(u,v) \qquad (7.34)$$

$$\mathcal{F}[f(ax,by)] = \frac{1}{|ab|} \, F\left(\frac{u}{a}, \frac{v}{b}\right) \qquad (7.35)$$

Equation (7.35) can be easily proved by direct substitution in Eq. (7.23) of ax, by, u/a, and v/y, respectively, for x, y, u, and v.

Periodicity and Conjugate Symmetry. It can be easily proved by substituting u + N for u or v + N for v in Eq. (7.23) that the Fourier transform and the inverse Fourier transform are periodic and have a period of N. Thus we have

$$F(u,v) = F(u + N,v) = F(u,v + N) = F(u + N,v + N) \qquad (7.36)$$

and also

$$f(x,y) = f(x + N, y) = f(x, y + N) = f(x + N, y + N) \qquad (7.37)$$

By the similar method, the conjugate symmetric property of the Fourier transform can also be proved such that

$$F(u,v) = F^*(-u,-v)$$

By the same token, the magnitude plot of the Fourier transform is symmetrical to the origin; or

$$|F(u,v)| = |F(-u,-v)| \qquad (7.38)$$

Rotation invariant. If polar coordinates (ρ,θ) and (r,ϕ) are introduced for the rectangular coordinate, the image function and its transform become $f(\rho,\theta)$ and $F(r,\phi)$, respectively. It can be easily shown by direct substitution into the Fourier transform pair that

$$\mathcal{F}\{f(\rho,\theta + \Delta\theta)\} = F(r, \phi + \Delta\phi) \qquad (7.39)$$

where $\Delta \phi = \Delta \theta$. That is, when the image function $f(x,y)$ is rotated by $\Delta \theta$, its Fourier transform is also rotated by the same angle $\Delta \theta$ (i.e., $\Delta \phi = \Delta \theta$). In other words, the same angle rotations occur in the spatial and the transform domains.

One more thing we would like to add about the Fourier transform is that most of information about the image object in the spatial domain concentrates at the central part of the spectrum. The intensity plot of the spectrum contains no positional information about the object, since the phase angle information is discarded in that plot. Complete reconstruction of the original image can be obtained when the real and imaginary spectrum data are included. The concentration of the spatial image information content in the Fourier spectrum is discussed in Sec. 7.4.3.

Correlation and Convolution. In the processing of images by the Fourier transform technique, correlation and convolution are the important operations. Emphasis on the clarification of the difference between them will be the main subject of this section. Definitions given hereafter are valid only for deterministic functions. Correlation of two continuous functions $f(x)$ and $g(x)$ is defined by

$$f(x) \circ g(x) = \int_{-\infty}^{\infty} f(\alpha)g(x + \alpha) \, d\alpha \qquad (7.40)$$

where α is a dummy variable of integration. The correlation is called *autocorrelation* if $f(x) = g(x)$, and *cross-correlation* if $f(x) \neq g(x)$. Convolution, by definition, is

$$f(x) * g(x) = \int_{-\infty}^{\infty} f(\alpha)g(x - \alpha) \, d\alpha \qquad (7.41)$$

The forms for correlation and convolution are similar, the only difference between them being the following. In convolution, $g(\alpha)$ is first folded about the vertical axis and then displaced by x to obtain a function $g(x - \alpha)$. This function is then multiplied by $f(\alpha)$ and integrated from $-\infty$ to ∞ for each value of displacement x to obtain the convolution (see Fig. 7.4, which is self-explanatory). In correlation, $g(\alpha)$ is not folded about the vertical axis and is directly displaced by x to obtain $g(x + \alpha)$. Figure 7.4e indicates the integral $\int f(\alpha)g(x + \alpha) \, d\alpha$ on the shaded areas shown in part (c), which is the correlation of $f(x)$ and $g(x)$; part (f) indicates the integral $\int f(\alpha)g(x - \alpha) \, d\alpha$ over the shaded area in part (d), which is the convolution of $f(x)$ and $g(x)$.

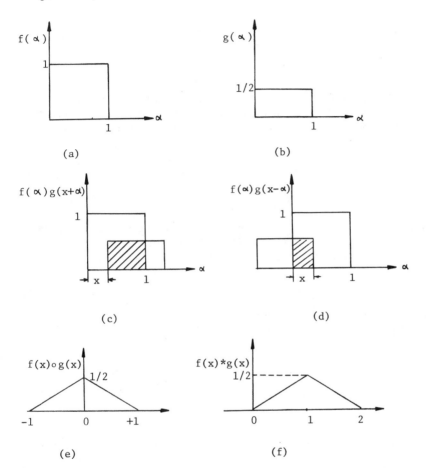

FIG. 7.4 Graphical illustration of the difference between correlation and convolution.

By the same definition, the correlation for a two-dimensional case can then be expressed as

$$f(x,y) \circ g(x,y) = \int_{-\infty}^{\infty} \int f(\alpha, \beta) g(x + \alpha, y + \beta) \, d\alpha \, d\beta \qquad (7.42)$$

After discretization, the correlation between $f(x,y)$ and $g(x,y)$ becomes

$$f(x,y) \circ g(x,y) = \sum_{m=0}^{N-1} \sum_{n=0}^{N-1} f(m,n)g(x + m, y + n)$$

for $x = 0, 1, \ldots, N - 1; \quad y = 0, 1, \ldots, N - 1$ \hfill (7.43)

If the Fourier transform of $f(x,y)$ is $F(u,v)$ and that of $g(x,y)$ is $G(u,v)$, then the Fourier transform of the correlation of two functions $f(x,y)$ and $g(x,y)$ is the product of their Fourier transforms with one of them conjugated. Thus

$$f(x,y) \circ g(x,y) \iff F(u,v)G^*(u,v) \hfill (7.44)$$

which indicates that the inverse transform of $F(u,v)G^*(u,v)$ gives the correlation of the two functions in the (x,y) domain. An analogous result is that the Fourier transform of the product of two functions $f(x,y)$ and $g(x,y)$ with one of them conjugated is the correlation of their Fourier transforms, and is formally stated as

$$f(x,y)g^*(x,y) \iff F(u,v) \circ G(u,v) \hfill (7.45)$$

where * represents the complex conjugate. These two results together constitute the correlation theorem. Similarly, we can derive the convolution theorem as follows:

$$f(x,y) * g(x,y) \iff F(u,v)G(u,v) \hfill (7.46)$$

and

$$f(x,y)g(x,y) \iff F(u,v) * G(u,v) \hfill (7.47)$$

This states that the Fourier transform of the convolution of two functions is equal to the product of the Fourier transforms of the two functions, and conversely the Fourier transform of a product of two functions is equal to the convolution of the Fourier transforms of the two functions. These two relations constitute the convolution theorem. This theorem is very useful in that complicated integration in the spatial domain can be completed by comparatively simpler multiplication in the Fourier domain.

One of the principal applications of correlation in image processing is in the area of template matching. The correlations of the unknown with those images of known origin are computed and the largest correlation indicates the closest match. This is sometimes called the method of maximum likelihood.

7.3 SAMPLING

In digital image processing by Fourier transformation, an image function is first sampled into uniformly spaced discrete values in the spatial domain and is then Fourier transformed. After the processing is completed in the Fourier domain, the results are converted back to the spatial image by inverse transformation. What interests us most is the relations between the sampling conditions and the recovered image from the set of sampled values. Let us start with the one-dimensional case.

7.3.1 One-Dimensional Functions

If we have a function f(x) which is the envelope of a string of impulses on Fig. 7.5a, we have a corresponding transform F(u) as shown in Fig. 7.5b. The string of impulses shown in part (a) is actually a

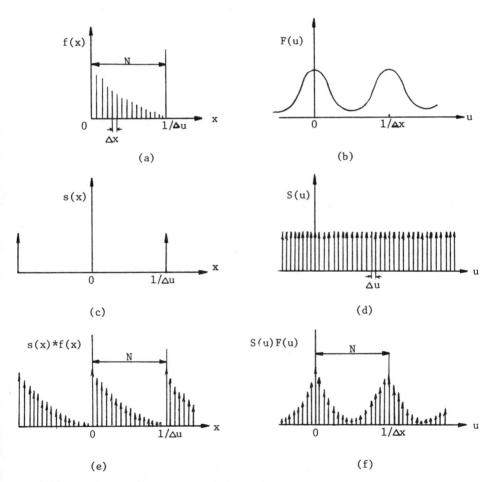

FIG. 7.5 Sampling on a one-dimensional function.

sampled version of f(x) with train of impulses Δx apart or

$$f(x) \sum_{k=-\infty}^{\infty} \delta(x - k \Delta x) = \sum_{k=-\infty}^{\infty} f(k \Delta x)\delta(x - k \Delta x) \qquad (7.48)$$

F(u) convoluted at interval $1/\Delta x$ is shown in part (b). This is because the Fourier transform of a string of impulses will be another string of impulses a distance $1/\Delta x$ unit apart, where Δx is the distance between impulses of the original string.

Similarly, if we sample F(u) with an impulse train S(u) Δu units apart between impulses, we will have f(x) $\Sigma \ \delta(x - k \Delta x)$ convoluted at interval $1/\Delta u$ and periodic with period $1/\Delta u$ in the spatial domain. If N samples of f(x) and F(u) are taken, and the spacings of Δu are selected such that the interval $1/\Delta u$ just covers N samples in the x domain and the interval $1/\Delta x$ just covers N samples in the frequency domain, then

$$\frac{1}{\Delta u} = N \Delta x \qquad (7.49)$$

or

$$\Delta u = \frac{1}{N} \frac{1}{\Delta x} \qquad (7.50)$$

7.3.2 Two-Dimensional Functions

For the two-dimensional case, the image function is f(x,y). The Fourier transform pair is

$$F(u,v) = \frac{1}{MN} \sum_{x=0}^{M-1} \sum_{y=0}^{N-1} f(x,y)e^{-j2\pi(ux/M+vy/N)}$$

$$\text{for } u = 0, 1, \ldots, M - 1; \quad v = 0, 1, \ldots, N - 1 \qquad (7.51)$$

and

$$f(x,y) = \sum_{u=0}^{M-1} \sum_{v=0}^{N-1} F(u,v)e^{j2\pi(ux/M+vy/N)}$$

$$\text{for } x = 0, 1, \ldots, M - 1; \quad y = 0, 1, \ldots, N - 1 \qquad (7.52)$$

Let Δx and Δy be the separations of strings in the spatial domains, and Δu and Δv be those in the frequency domain;

$$\Delta u = \frac{1}{M \Delta x}$$

$$\qquad (7.53)$$

$$\Delta v = \frac{1}{N \Delta y}$$

can be worked out for this two-dimensional DFT by the analogous analysis developed for the one-dimensional case. With these relation-

ships, the image function and its two-dimensional Fourier transform
will be periodic with M × N uniformly spaced values. For N × N square
array samples, that is, M = N, we have

$$\Delta u = \frac{1}{N \, \Delta x}$$

$$\Delta v = \frac{1}{N \, \Delta y} \qquad\qquad (7.54)$$

These relationships between the sample separations guarantee that the
two-dimensional period defined by $1/\Delta u \times 1/\Delta v$ will just be covered by
N × N uniformly spaced samples in the spatial domain and the period
defined by $1/\Delta x \times 1/\Delta y$ will be covered by N × N samples in the frequency domain. It is noted that the constant multiplicative terms may
be grouped arbitrarily. If they are regrouped this way such that
$NF(u,v) \Rightarrow F(u,v)$, the Fourier transform pair will have the following
form:

$$F(u,v) = \frac{1}{N} \sum_{x=0}^{N-1} \sum_{y=0}^{N-1} f(x,y) e^{-j2\pi(ux+vy)/N}$$

$$u, \, v = 0, \, 1, \, \ldots, \, N - 1 \qquad\qquad (7.55)$$

$$f(x,y) = \frac{1}{N} \sum_{u=0}^{N-1} \sum_{v=0}^{N-1} F(u,v) e^{j2\pi(ux+vy)/N}$$

$$x, \, y = 0, \, 1, \, \ldots, \, N - 1 \qquad\qquad (7.56)$$

The notions of impulse sheets and two-dimensional impulses suggested by Lendaris et al. (1970) will be introduced for the discussion
of two-dimensional functions. An impulse sheet is defined such that
it has an infinitive length in one direction and its cross section has
the usual delta-function properties. The cross section of the impulse
sheet is presumed to remain the same along the sheet's entire length.
The intersection of two impulse sheets results in a two-dimensional
impulse located at their intersection.

The impulse sheets and two-dimensional impulses have the
following properties:

1. The two-dimensional transform of an impulse sheet is an
    impulse sheet centered at the origin and in the direction or-
    thogonal to the direction of the original impulse sheet.
2. The two-dimensional Fourier transform of an infinitive array
    of uniformly spaced parallel impulse sheets is an infinite
    string of impulses along the direction orthogonal to the impulse
    sheet direction, with a spacing inversely proportional to the

impulse sheet separation, and with one of the impulses located
at the origin.

3. Conversely, the two-dimensional Fourier transform of a string
    of uniformly spaced impulses is an array of parallel impulse
    sheets whose direction is orthogonal to the impulse string,
    whose separation is inversely proportional to the impulse
    separation, and one of whose impulse sheets goes through
    the origin.

4. The two-dimensional Fourier transform of an infinite lattice-
    like array of impulses is an infinite lattice-like array of im-
    pulses whose dimensions are inversely related to those of the
    original lattice, with an impulse at the origin.

5. When convolving a function with an array of impulses, the
    function is simply replicated at the locations of each of the
    impulses.

6. The convolution theorem as derived for one-dimensional
    functions also holds for two-dimensional functions.

Relatively Large Aperture

If the image consists of an array of uniformly spaced parallel straight
lines with a scan circular aperture, the Fourier transform of the combination will be the convolution of their respective Fourier transforms.
Thus

$\mathcal{F}$[(a straight line $*$ a string of impulses)(circular aperture
function)]

$= \mathcal{F}\ (\cdot) * \mathcal{F}\ (\cdot)$

where the first $\mathcal{F}\ (\cdot)$ is the Fourier transform of the first set of parentheses, while the second $\mathcal{F}\ (\cdot)$ is the Fourier transform of the second set of parentheses. The result will be a string of impulses with
a separation equal to the reciprocal of the spacing of the lines and in
a direction perpendicular to these parallel lines. If the aperture is
relatively large with respect to the spacing of the parallel lines, the
separation of the string of impulses will be large, as shown in Fig.
7.6b. The airy disk is the Fourier transform of the circular aperture,
and is replicated on each impulse.

Figure 7.7 shows another example, which consists of an array
of rectangles with a rectangular aperture. The Fourier transform of
this image will be the Fourier transform of the combination

[(a rectangle $*$ array of impulses)](rectangular aperture function)

The result of Fourier transform of this array of rectangles (i.e., those
inside the brackets) will be the product of Fourier transform of the

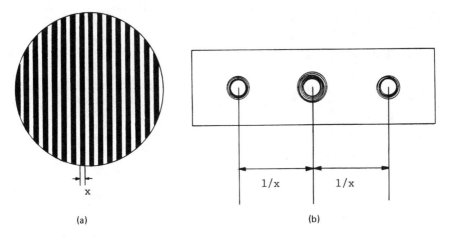

FIG. 7.6 Fourier transform of uniformly spaced parallel straight lines with a scan circular aperture: (a) uniformly spaced parallel straight lines; (b) Fourier transform of (a).

rectangle shown in Fig. 7.7b and the Fourier transform of the array of impulses. The Fourier transform of the array of impulses is another array of impulses. Multiplication of the Fourier transform of the rectangle with an array of impulses gives a "sampling" Fourier transform of the rectangle (Fig. 7.7c).

By the convolution theorem, the product of the rectangular aperture function and the array of rectangles in the spatial domain will give a convolution of their respective spectra. Therefore, the Fourier transform of the rectangular aperture will be replicated at each sampling point of the Fourier transform in Fig. 7.7c.

Relatively Small Aperture

Let us take the example shown in Fig. 7.8. This array of six squares can be viewed as the result of convolving one square with an infinite array of impulses, with the result multiplied by an aperture to allow only the six squares to appear; that is,

[(a square ∗ infinite array of impulses)](rectantular aperture)

According to the mathematical operation above, its corresponding Fourier transform will then be

[(FT of the square)(FT of the impulse array)] ∗ (FT of rectangular aperture)

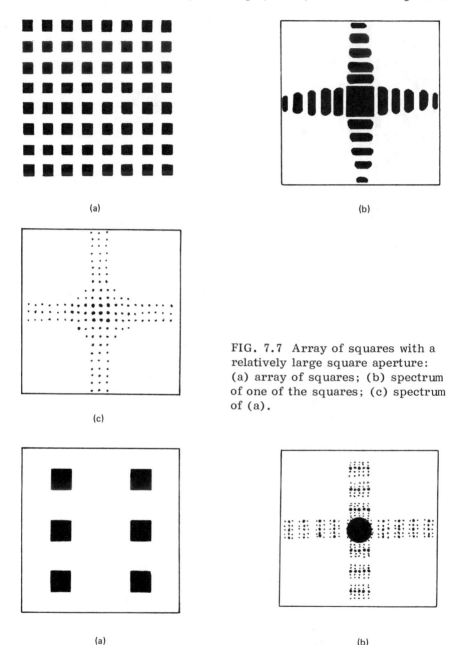

(a)

(b)

(c)

FIG. 7.7 Array of squares with a relatively large square aperture: (a) array of squares; (b) spectrum of one of the squares; (c) spectrum of (a).

(a)

(b)

FIG. 7.8 Array of squares with relatively small aperture: (a) array of squares; (b) spectrum.

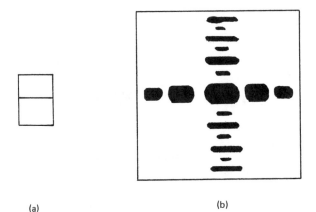

(a)   (b)

FIG. 7.9 Primitive of a Chinese character and its spectrum: (a) primitive; (b) central portion of the spectrum of (a).

Since the impulse spacing in the spatial domain is relatively large with respect to the aperture, the Fourier transform of this array of impulses (which turns out to be another array of impulses) will have relatively narrow spacing, thus giving a sampled version of the Fourier transform of the square. Convolution of this multiplication with the Fourier transform of the rectangular aperture will have the Fourier transform of the aperture replicated at each sample point.

In extracting primitives of Chinese characters by the Fourier transform technique, a similar problem appears. Figure 7.9a shows a

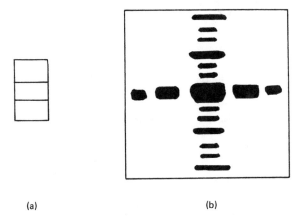

(a)   (b)

FIG. 7.10 Primitive of a Chinese character and its spectrum: (a) primitive; (b) central portion of the spectrum of (a).

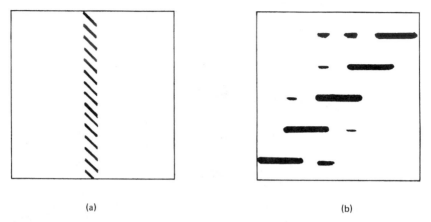

(a)   (b)

FIG. 7.11 Parked cars (staggered rectangular objects): (a) parked cars; (b) central portion of the spectrum of (a).

primitive consisting of two parallel bars in a vertical direction and three bars in a horizontal direction. Figure 7.9b shows the central portion of the Fourier transform of part (a). Another primitive, different by one more bar in the vertical direction from that shown in part (a), is shown in Fig. 7.10a. Because of the difference in the number of bars in the vertical direction, the difference in the Fourier transforms can also be noted.

Figure 7.11a shows a parking lot. For the same reason as before, this can be viewed as a small rectangle convoluted with an infinite string of uniformly spaced impulses and multiplied by the scan aperture so as to cause only a small number of rectangles to appear. Figure 7.11b shows the central portion of the Fourier transform of Fig. 7.11a.

7.3.3 Applications

Sampling is a good tool to use to reduce the amount of information to be processed. The Fourier spectrum is particularly well suited to sampling because of the following features:

1. Most of the information from the original imagery will be on the central portion of the spectrum. From the translation property of the Fourier transform,

$$f(x - x_0, y - y_0) \iff F(u,v)e^{-j2\pi(ux_0+vy_0)/N} \qquad (7.57)$$

It is interesting to note that a shift in $f(x,y)$ does not affect the magnitude of its transform. A linear object in the original imagery gives rise to a spectrum along a line centered on the central axis.

Similarly, a circular object gives rise to a spectrum as concentric annular rings centered on the central axes. A lattice-like object gives rise to a lattice-like spectrum in the diffraction patterns. However, no reference point like this exists in the original spatial image. Scanning and processing of the whole area are then necessary to obtain the object information in the image.

2. From the linearity property of Fourier transform as described by

$$\mathcal{F}[af_1(x,y) + f_2(x, y)] = aF_1(u,v) + F_2(u,v) \qquad (7.58)$$

where $F_1(u,v) = \mathcal{F}[f_1(x,y)]$ and $F_2(u,v) = \mathcal{F}[f_2(x,y)]$, the Fourier spectrum can be well interpreted by superposition of component spectra from their corresponding separable spatial image functions.

Several sampling devices were suggested by Lendaris et al. (1970) to measure the amount of light energy falling within specified areas of the Fourier spectrum. With an annular-ring sampling device as shown in Fig. 7.12a, the total light energy of the Fourier spectrum measured along a circle centered on the optical axis corresponds to one frequency in all directions. With a set of annular-ring sampling windows, a spatial frequency profile of the contents of the scan area can be obtained simultaneously. The device can be used to detect the regularity.

Figure 7.12b shows another sampling device (a wedge-shaped sampling window) in which the light energy of the spectrum along a radial line (which corresponds to a single direction in the spectrum)

(a)   (b)

FIG. 7.12 Sampling devices: (a) annular-ring sampling device; (b) wedge-shaped sampling device.

can be measured, and gives a direction profile of the contents of the scan area simultaneously. This device can be used to find the principal directions. These sample signatures, obtained either from the annular ring or from a wedge-shaped sampling device (or from both) are useful for pattern recognition.

7.4 FAST FOURIER TRANSFORM

The Fourier transformation technique is a very effective tool, although a great deal of computation is needed to carry out these transformations. This makes the Fourier transformation technique impractical unless the computation can be simplified.

7.4.1 DFT of a Two-Dimensional Image Computed as a Two-Step One-Dimensional DFT

The separability of the kernel property can be used in the simplification of the transformation process. Rewrite Eqs. (7.32) and (7.33) as follows:

$$F(u,v) = \mathcal{F}_y \{ \mathcal{F}_x [f(x,y)] \} \qquad (7.59)$$

and

$$F(u,v) = \mathcal{F}_x \{ \mathcal{F}_y [f(x,y)] \} \qquad (7.60)$$

In other words, the Fourier transformation operation on the image function $f(x,y)$ can be performed in two steps: first, transform the two-dimensional image function column-wise [i.e., perform one-dimensional transform along each column of the image function $f(x,y)$], and then transform the results row-wise (i.e., perform one-dimensional transform along each row of the resulting spectrum), as indicated by Eq. (7.59). A different order of transformation can also be taken: transform $f(x,y)$ row-wise first, and then transform the resulting spectrum column-wise, as indicated by Eq. (7.60).

Similarly, the inverse Fourier transform can also be performed in two steps, such as

$$f(x,y) = \mathcal{F}_{u,v}^{-1} [F(u,v)] = \mathcal{F}_u^{-1} \{ \mathcal{F}_v^{-1} [F(u,v)] \} \qquad (7.61)$$

and

$$f(x,y) = \mathcal{F}_v^{-1} \{ \mathcal{F}_u^{-1} [F(u,v)] \} \qquad (7.62)$$

The complex conjugate properties in the arithmetic operation can also be used to simplify the transformation process. The conjugate of [f(x,y)]* can be written according to Eq. (7.56) as

$$[f(x,y)]^* = \frac{1}{N}\left[\sum_{u=0}^{N-1}\sum_{v=0}^{N-1} F(u,v)e^{j2\pi(ux+vy)/N}\right]^* \qquad (7.63)$$

where $e^{j2\pi(ux+vy)/N}$ is the inverse transformation kernel. Equation (7.63) can further be written as

$$[f(x,y)]^* = \frac{1}{N}\sum_{u=0}^{N-1}\sum_{v=0}^{N-1} F(u,v)^* e^{-j2\pi(ux+vy)/N} \qquad (7.64)$$

from which it is interesting to know that the inverse transformation kernel in Eq. (7.63) is converted to a forward transformation kernel in Eq. (7.64). That is, it is possible to use the same forward transformation kernel to do the inverse transformation. For a real function, where [f(x,y)]* = f(x,y), Eq. (7.64) then becomes

$$f(x,y) = \frac{1}{N}\sum_{u=0}^{N-1}\sum_{v=0}^{N-1} F^*(u,v)e^{-j2\pi(ux+vy)/N} \qquad (7.65)$$

Comparison with Eq. (7.55) shows that the same forward transformation algorithm can be used to do the inverse transform, as far as the Fourier transform F(u,v) is conjugated to F*(u,v). Arguments similar to those used in Eqs. (7.59) and (7.60) are also valid for Eq. (7.65); thus

$$f(x,y) = \mathcal{F}_{u,v}[F^*(u,v)] = \mathcal{F}_u\{\mathcal{F}_v[F^*(u,v)]\} \qquad (7.66)$$

Note that unlike Eq. (7.62), $\mathcal{F}_u$ and $\mathcal{F}_v$ in Eq. (7.66) are forward Fourier transforms. A conclusion can then be drawn that an inverse discrete Fourier transform may be computed as the discrete Fourier transform of the conjugate, and can also be computed as a two-step one-dimensional discrete Fourier transform. One-dimensional discrete Fourier transform will then be the nucleus of the discrete Fourier transform and the inverse discrete Fourier transform.

7.4.2 Method of Successive Doubling

As discussed in previous sections, a two-dimensional Fourier transform can be separated into two computational steps, each of which is a one-dimensional Fourier transform, and an inverse two-dimensional Fourier transform may be computed as the discrete Fourier transform of the conjugate and can also be computed as a two-step one-dimensional discrete Fourier transform. Thus, a one-dimensional discrete Fourier transform will be the nucleus of the computation and more effort should be concentrated on its algorithm design to make it more effective.

The discrete Fourier transform for one-dimensional function is rewritten as follows from Eq. (7.2):

$$F(u) = \frac{1}{N} \sum_{x=0}^{N-1} f(x)e^{-j2\pi ux/N} \qquad u = 0, 1, \ldots, N-1 \qquad (7.67)$$

Regrouping in the same manner as in Eqs. (7.55) and (7.56) the constant multiplication factor for the convenience of analysis and letting $W = e^{-j2\pi/N}$, then the discrete Fourier transform becomes

$$F(u) = \sum_{x=0}^{N-1} f(x)W^{ux} \qquad \text{for } u = 0, 1, \ldots, N-1 \qquad (7.68)$$

giving a system of N simultaneous equations, corresponding to N different values of u. It can be easily seen from Eq. (7.68) that N complex multiplications and N complex additions are needed for each equation, or a total of $2N^2$ complex arithmetic operations for the whole array (or a total of N^2 operations when f(x) is a real function). When N becomes large, the number of computations involved in the Fourier transform is terribly great. Hence, computation must be simplified to make the transformation technique practical. It became so only when the fast Fourier transform (FFT) was suggested in 1965. The fundamental principle in FFT algorithm is based on the decomposition of DFT computation of a sequence of length N into successively smaller DFTs.

Assume that $N = 2^L$, where L is a positive integer, Eq. (7.68) can then be broken into two parts:

$$F(u) = \sum_{\substack{\text{even} \\ x}} f(x)W_N^{ux} + \sum_{\substack{\text{odd} \\ x}} f(x)W_N^{ux} \qquad \text{for } u = 0, 1, \ldots, N-1$$
$$(7.69)$$

where $W_N = e^{-j2\pi/N}$. There are still N terms (or a sequence of N terms, in each equation of the above system. Equation (7.69) can, in turn, be put in the following form with r equal to a positive integer:

$$F(u) = \sum_{r=0}^{(N/2)-1} f(2r)W_N^{2ru} + \sum_{r=0}^{(N/2)-1} f(2r+1)W_N^{(2r+1)u} \qquad (7.70)$$

or

$$F(u) = \sum_{r=0}^{(N/2)-1} f(2r)(W_N^2)^{ru} + W_N^u \sum_{r=0}^{(N/2)-1} f(2r+1)(W_N^2)^{ru}$$

$$(7.71)$$

The first summation on the right-hand side consists of a sequence of $N/2$ terms. Note from the definition of W_N that

$$W_N^2 = (e^{-j2\pi/N})^2 = e^{-j2\pi/(N/2)} = W_{N/2} \qquad (7.72)$$

Use $W_{N/2}$ as the kernal for the sequence of $N/2$ terms; then we have

$$F(u) = \sum_{r=0}^{(N/2)-1} f(2r)(W_{N/2})^{ru} + W_N^u \sum_{r=0}^{(N/2)-1} f(2r+1)(W_{N/2})^{ru}$$

$$(7.73)$$

or

$$F(u) = G(u) + W_N^u H(u) \qquad (7.74)$$

where

$$G(u) = \sum_{r=0}^{(N/2)-1} f(2r)(W_{N/2})^{ru} \qquad \text{and}$$

$$H(u) = \sum_{r=0}^{(N/2)-1} f(2r+1)(W_{N/2})^{ru}$$

are two $(N/2)$-point discrete Fourier transforms. Note also that

$$(W_{N/2})^{u+N/2} = (W_{N/2})^u \qquad (7.75)$$

Both $G(u)$ and $H(u)$ are periodic in u with a period of $N/2$. Careful analysis of Eqs. (7.69) through (7.75) reveals some interesting properties of these expressions. It is noted in Eqs. (7.73) and (7.74) that an N-point discrete transform can be computed by dividing the original expression into two parts. Each part corresponds to a $(N/2)$-point discrete transform computation. Obviously, the computation time required for $(N/2)$-point DFT will be more greatly reduced than that for N-point DFT.

By continuing this analysis, the (N/2)-point discrete transform can be obtained by computing two (N/4)-point discrete transforms, and so on, for any N that is equal to an integer power of 2. The implementation of these equations constitutes the successive-doubling FFT algorithm.

The implementation (Sze, 1979) of Eq. (7.73) is shown in Fig. 7.13 for N = 8. Inputs to the upper (N/2)-point DFT block are f(x)'s for even values of x, while those for the lower (N/2)-point DFT block are f(x)'s for odd values of x. Substitution of values 0, 1, 2, and 3 for u in

$$G(u) = \sum_{r=0}^{(N/2)-1} f(2r)(W_{N/2})^{ru} \qquad (7.76)$$

and

$$H(u) = \sum_{r=0}^{(N/2)-1} f(2r+1)(W_{N/2})^{ru} \qquad (7.77)$$

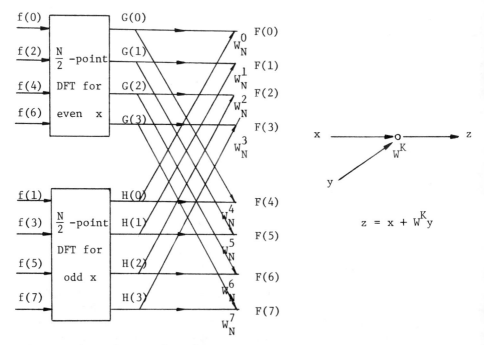

FIG. 7.13 Implementation of Eq. (7.73) for N = 8.

gives $G(0), \ldots, G(3)$ and $H(0), \ldots, H(3)$; which combine according to the signal flow graph indicated in the figure to give the Fourier transforms $F(u)$'s, $u = 0, 1, \ldots, 7$. $W_N^0, \ldots, W_N^7$ on the graph indicate the multiplying factors on $H(u)$'s, $u = 0, 1, \ldots, 3$, needed in Eq. (7.74). Note that $G(u)$ and $H(u)$ are periodic in u with a period of $N/2$, which is 4 in this case [i.e., $G(4) = G(0)$; $G(5) = G(1)$; $H(4) = H(0)$; $H(5) = H(1)$; etc.]. Thus $F(7) = G(7) + W_N^7 H(7) = G(3) + W_N^7 H(3)$. With this doubling algorithm, the number of complex operations needed for the eight-point DFT is reduced. The total number of complex operations required before using the doubling algorithm is 8^2 or 64, whereas that needed after the doubling algorithm is used is $8 + 2(8/2)^2$ or 40, where $(8/2)^2$ is the number of mathematical operations needed for each of the $(8/2)$-point DFT's, and 8 (the first term in the expression) is the number of addition operations needed. Replacing $f(2r)$ by $g(r)$ and letting $r = 2\ell$ in Eq. (7.73), we then separate $G(u)$ into two parts, one for even r's and the other one for odd r's. We then have

$$G(u) = G_1(u) + W_{N/2}^u G_2(u) \tag{7.78}$$

where

$$G_1(u) = \sum_{\ell=0}^{(N/4)-1} g(2\ell) W_{N/4}^{\ell u} \tag{7.79}$$

represents that part of $G(u)$ for even values of r, and

$$G_2(u) = \sum_{\ell=0}^{(N/4)-1} g(2\ell + 1) W_{N/4}^{\ell u} \tag{7.80}$$

represents that part of $G(u)$ for odd values of r. $G_1(u)$ and $G_2(u)$ are periodic with period of $N/4$. Similarly we have

$$H(u) = H_1(u) + W_{N/2}^u H_2(u) \tag{7.81}$$

where $H_1(u)$ represents that part of $H(u)$ for even values of r, and $H_2(u)$ for odd values of r. Again, $H_1(u)$ and $H_2(u)$ are periodic with a period of $N/4$. Then the $(N/2)$-point DFT is decomposed as shown in Fig. 7.14. The upper $(N/4)$-point DFT block implements Eq. (7.79) and the lower $(N/4)$-point DFT block implements Eq. (7.80).

The decomposition of the DFT computation keeps on going until $(N/2^{L-1})$-point DFT becomes a two-point DFT. For the case we have now, that is, $N = 8$, $(N/4)$-point DFT is a two-point DFT, where

$$G_1(0) = f(0) + f(4)W_{N/4}^0 = f(0) + f(4)$$

$$G_1(1) = f(0) + W_{N/4}^1 f(4) = f(0) - f(4)$$

(7.82)

The complete eight-point DFT (or FFT for the case N = 8) is implemented as shown in Fig. 7.15.

By means of the successive-doubling algorithm, the total number of complex operations changes from the original N^2 to $N + 2(N/2)^2$, and then to $N + 2[N/2 + 2(N/4)^2]$, and so on, depending on the number of stages into which the N-point DFT can be decomposed. If N is large and equal to 2^L, then the number of stages is L, and the number of complex operations changes from N^2 to $N + N + \cdots + N = N \times L = N \log_2 N$. For the example just given, the total number of complex operations will be $N \times L = 8 \times 3 = 24$.

To maintain the structure of this algorithm, the inputs to the DFT block must be arranged in the order required for successive application of Eq. (7.73). For the FFT computation of an eight-point function {f(0), f(1), ..., f(7)}, inputs with even arguments f(0), f(2), f(4), f(6) are used for the upper (N/2)-point DFT (four-point DFT in this case), while those with odd arguments f(1), f(3), f(5), f(7) are used for the lower four-point DFT. Each four-point transform is computed as two 2-point transforms. We must divide the first set of inputs into its even part {f(0), f(4)} and odd part {f(2), f(6)}, and divide the second set of inputs into {f(1), f(5)} as the even part and {f(3), f(7)} as the odd part. That is to say, we must arrange the inputs in the order {f(0), f(4), f(2), f(6), f(1), f(5), f(3), f(7)} for the successive-

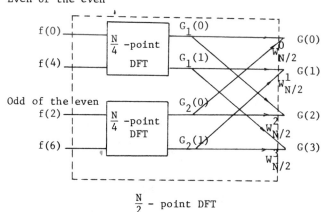

FIG. 7.14 Implementation of Eq. (7.78) for N = 8.

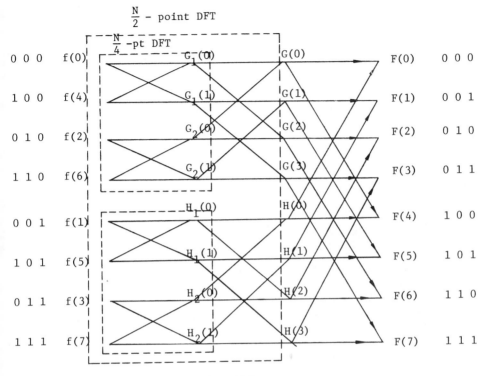

FIG. 7.15 Complete eight-point DFT.

doubling algorithm of an eight-point function as shown in Fig. 7.16. It is not difficult to note that the input and output are related by a "bit-reversal" order, as shown in Table 7.1. Note that in Fig. 7.15, $W_N = e^{-j2\pi/N}$ and $N = 8$; we then have $W_N^0 = 1$, $W_N^4 = -1$, $W_N^5 = -W_N^1$, $W_N^6 = -W_N^2$, and $W_N^7 = -W_N^3$. Utilizing these relations, Fig. 7.17 results.

TABLE 7.1 Example of Bit Reversal and Reordering of Inputs for FFT Algorithm

Input order	Argument	Binary-coded			Bit-reversal binary-coded			New argument	Output order
$f(0)$	0	0	0	0	0	0	0	0	$F(0)$
$f(4)$	4	1	0	0	0	0	1	1	$F(1)$
$f(2)$	2	0	1	0	0	1	0	2	$F(2)$
$f(6)$	6	1	1	0	0	1	1	3	$F(3)$

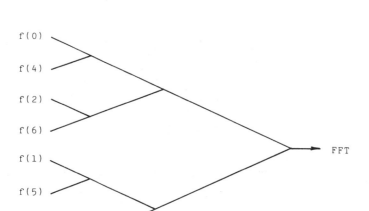

FIG. 7.16 Reordering of inputs for successive doubling method.

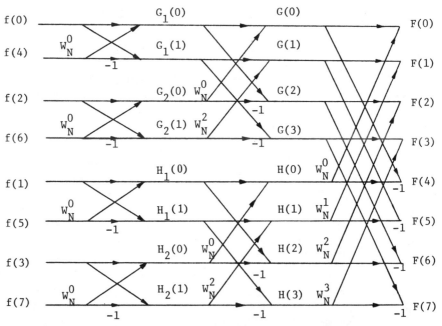

FIG. 7.17 "Bit-reversal" order relationship between input and output in the FFT algorithm.

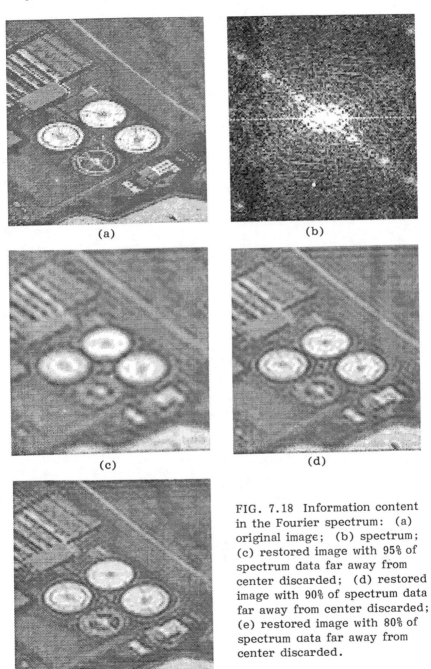

(a)

(b)

(c)

(d)

(e)

FIG. 7.18 Information content in the Fourier spectrum: (a) original image; (b) spectrum; (c) restored image with 95% of spectrum data far away from center discarded; (d) restored image with 90% of spectrum data far away from center discarded; (e) restored image with 80% of spectrum aata far away from center discarded.

TABLE 7.2   Computational Advantage Obtained for Various Values of N

N	L	Conventional two-dimensional FFT N^4	Two-dimensional FFT $N^2 \log_2 N$	Computational advantage $N^2/\log_2 N$
2	1	16	4	4.00
'4	2	256	32	8.00
8	3	4.096	192	21.33
16	4	6.554×10^4	1,024	64.00
32	5	1.049×10^6	5,120	204.8
64	6	1.678×10^7	2.458×10^4	682.67
128	7	2.684×10^8	1.147×10^5	2,341.0
256	8	4.295×10^9	5.243×10^5	8,192.0
512	9	6.872×10^{10}	2.359×10^6	2.913×10^4
1,024	10	1.100×10^{12}	1.049×10^7	1.049×10^5

Owing to the fact that $W_N^{r+N/2} = -W_N^r$, the number of complex multiplications needed for the successive-doubling FFT computational configuration would be further reduced by a factor of 2. Therefore, the total number of complex operations needed for one-dimensional DFT would be $(N \log_2 N)/2$. For a two-dimensional $N \times N$ image function $f(x,y)$, we need $(N \log_2 N)/2$ computational operations for one value of u, and therefore $(N \times N \log_2 N)/2$ for N values of u. By the same reasoning, we need $(N \times N \log_2 N)/2$ operations for N values of v. The total number of complex operations will then be $N^2 \log_2 N$. But if the two-dimensional Fourier transform is evaluated directly, $(N^2)(N^2) = N^4$ complex operations will be required. Table 7.2 shows a comparison of N^4 versus $N^2 \log_2 N$ for various values of N.

Thus far, discussions have emphasized the forward FFT. As discussed in Sec. 7.4.1, the inverse transform can be performed using the same transformation algorithm as long as F(u,v) is conjugated to F*(u,v).

7.4.3   Image Information Content in the Fourier Spectrum: A Practical Example

As discussed in Sec. 7.3, every component object in a scan area has its own spectrum. All these spectra will be superimposed and centered on the central axes. It is interesting to know how many percentages of data taken from the central spectrum portion will be enough to preserve the image quality. No general answer can be given, since this is highly problem dependent. For some cases the picture quality is the most important requirement and therefore, a larger percentage of the spectrum data should be used in the processing to preserve both the low-frequency and all the high-frequency components of the image. But in other cases, the processing speed will be the first criterion, and even some sacrifice of image quality will be tolerable. Numerous examples can be enumerated, one of which is air reconnaissance. All we need is to search the desired objects within the scan area at a very fast speed and then focus our analysis on a smaller area to get more details. The first thing of concern is the speed; that is what is usually required in real-time processing or pseudo-real-time processing.

The complex image shown in Fig. (7.18a) has been Fourier transformed and different percentages (5%, 10%, and 20%) of its spectrum were taken to restore the images. The processing results are shown in Fig. 7.18. Part (a) is the original image, (b) is its spectrum, and (c), (d), and (e) show, respectively, the restored images when 95%, 90%, and 80% of the spectrum information far away from the center was discarded. It can be seen that lots of boundary information has been lost in the restored image shown in (c) (i.e., when 95% of the spectrum data was discarded in the restoration process). But restored images like those shown in part (e) are sometimes acceptable for applications where high processing speed is of primary concern.

7.5 OTHER IMAGE TRANSFORMS

The Fourier transform is just one of the transformation techniques frequently used in image processing. Other transformation techniques have also been shown to be very effective: Walsh transforms, Hadamard transforms, Karhunen-Loéve transforms, and so on. Like Fourier transforms, all of these transforms are reversible; that is, both forward and inverse transforms can be operated on functions that are continuous and integrable, thus making it possible to process an image in the transform domain.

7.5.1 Walsh Transform

If the function

$$g(x,u) = \frac{1}{N} \prod_{i=0}^{n-1} (-1)^{b_i(x)b_{n-1-i}(u)} \tag{7.83}$$

is used for the forward transform kernel in the generalized transformation equation (7.10), the transformation is known as the Walsh transform. Thus the Walsh transform of a function $f(x)$ is

u \ x	0	1	2	3	4	5	6	7	8	9	10	11	12	13	14	15
0	+	+	+	+	+	+	+	+	+	+	+	+	+	+	+	+
1	+	+	+	+	+	+	+	+	-	-	-	-	-	-	-	-
2	+	+	+	+	-	-	-	-	+	+	+	+	-	-	-	-
3	+	+	+	+	-	-	-	-	-	-	-	-	+	+	+	+
4	+	+	-	-	+	+	-	-	+	+	-	-	+	+	-	-
5	+	+	-	-	+	+	-	-	-	-	+	+	-	-	+	+
6	+	+	-	-	-	-	+	+	+	+	-	-	-	-	+	+
7	+	+	-	-	-	$(-)$	+	+	-	-	+	+	+	+	-	-
8	+	-	+	-	+	-	+	-	+	-	+	-	+	-	+	-
9	+	-	+	-	+	-	+	-	-	+	-	+	-	+	-	+
10	+	-	+	-	-	+	-	+	+	-	+	-	-	+	-	+
11	+	-	+	-	-	+	-	+	-	+	-	+	+	-	+	-
12	+	-	-	+	+	-	-	+	+	-	-	+	+	-	-	+
13	+	-	-	+	+	-	-	+	-	+	+	-	-	+	+	-
14	+	-	-	+	-	+	+	-	+	-	-	+	-	+	+	-
15	+	-	-	+	-	+	+	-	-	+	+	-	+	-	-	+

FIG. 7.19 Values of the Walsh transformation kernel for N = 16.

$$W(u) = \frac{1}{N} \sum_{x=0}^{N-1} f(x) \prod_{i=0}^{n-1} (-1)^{b_i(x) b_{n-1-i}(u)} \qquad (7.84)$$

where N is the number of samples and is assumed to be 2^n, with n
as a positive integer. $b_k(z)$ represents the kth bit in the binary
representation of z with the zeroth bit as the least significant one.
For example, if n = 4, z = 13 (1 1 0 1 in binary representation),
then $b_0(z) = 1$, $b_1(z) = 0$, $b_2(z) = 1$, and $b_3(z) = 1$. The kernel for
n = 4 is

$$g(x,u) = \frac{1}{N} \sum_{i=0}^{n-1} (-1)^{b_i(x) b_{n-1-i}(u)}$$

$$= \frac{1}{N} [(-1)^{[b_0(x)b_3(u)+b_1(x)b_2(u)+b_2(x)b_1(u)+b_3(x)b_0(u)]}$$

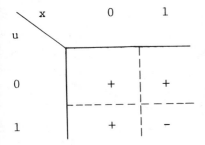

FIG. 7.20 Values of the Walsh transformation kernel for N = 2.

By substituting x = 5, u = 7 in the expression above, we obtain the value of the kernel for the circled entry in Fig. 7.19 as

$$g(5,7) = \frac{1}{N}[(-1)^{(1 \times 0 + 0 \times 1 + 1 \times 1 + 0 \times 1)}]$$

$$= \frac{1}{N}[(-1)^{1}] = -\frac{1}{N}$$

which is a negative value. It can be seen from Fig. 7.19 that the kernel is symmetrical and orthogonal, and therefore the inverse kernel h(x,u) is identical to the forward kernel, except for the constant multiplicative factor 1/N. Hence we have

$$h(x,u) = \prod_{i=0}^{n-1} (-1)^{b_i(x)b_{n-1-i}(u)} \tag{7.85}$$

The inverse Walsh transform is then

$$f(x) = \sum_{u=0}^{N-1} W(u) \prod_{i=0}^{n-1} (-1)^{b_i(x)b_{n-1-i}(u)} \tag{7.86}$$

Let us start from the smallest N (N = 2) and see how the array builds up with the Walsh transformation kernel. When N = 2 (or n = 1), Eq. (7.83) becomes

$$g(x,u) = \frac{1}{N}(-1)^{b_0(x)b_0(u)}$$

The simplest kernal in the Walsh transformation will be that as shown in Fig. 7.20. For N = 4 (or n = 2), Eq. (7.83) becomes

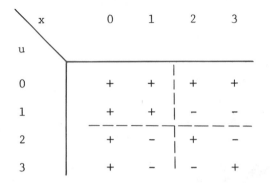

FIG. 7.21　Values of the Walsh transformation kernel for N = 4.

$$g(x,u) = \frac{1}{N}(-1)^{[b_0(x)b_1(u)+b_1(x)b_0(u)]}$$

The corresponding Walsh transformation kernel will be the array shown in Fig. 7.21. Following the same process of arithmetic substitution, the arrays formed for the Walsh transformation kernel for N = 8 (or n = 3) and for N = 16 (or n = 4) are shown in Figs. 7.22 and 7.19, respectively.

Extending the Walsh transformation as derived above for the two-dimensional case, we obtain the transformation kernel pair as follows:

u \ x	0	1	2	3	4	5	6	7
0	+	+	+	+	+	+	+	+
1	+	+	+	+	−	−	−	−
2	+	+	−	−	+	+	−	−
3	+	+	−	−	−	−	+	+
4	+	−	+	−	+	−	+	−
5	+	−	+	−	−	+	−	+
6	+	−	−	+	+	−	−	+
7	+	−	−	+	−	+	+	−

FIG. 7.22　Values of the Walsh transformation kernel for N = 8.

$$g(x,y;u,v) = \frac{1}{N} \prod_{i=0}^{n-1} (-1)^{[b_i(x)b_{n-1-i}(u)+b_i(y)b_{n-1-i}(v)]} \qquad (7.87)$$

and

$$h(x,y;u,v) = \frac{1}{N} \prod_{i=0}^{n-1} (-1)^{[b_i(x)b_{n-1-i}(u)+b_i(y)b_{n-1-i}(v)]} \qquad (7.88)$$

As discussed in Eq. (7.85), the same kernel can be used for both forward and inverse transformations, so we can then write the Walsh transform pair as follows:

$$W(u,v) = \frac{1}{N} \sum_{x=0}^{N-1} \sum_{y=0}^{N-1} f(x,y) \prod_{i=0}^{n-1} (-1)^{[b_i(x)b_{n-1-i}(u)+b_i(y)b_{n-1-i}(v)]}$$

$$(7.89)$$

and

$$f(x,y) = \frac{1}{N} \sum_{u=0}^{N-1} \sum_{v=0}^{N-1} W(u,v) \prod_{i=0}^{n-1} (-1)^{[b_i(x)b_{n-1-i}(u)+b_i(y)b_{n-1-i}(v)]}$$

$$(7.90)$$

Equations (7.89) and (7.90) demonstrate that one algorithm can be used for the computation of both the forward and inverse two-dimensional Walsh transforms. It is obvious from Eqs. (7.89) and (7.90) that the transformation kernels $g(x,y;u,v)$ and $h(x,y;u,v)$ are symmetrical and separable, or

$$g(x,y;u,v) = g_1(x,u)g_2(y,v) \qquad (7.91)$$

$$h(x,y;u,v) = h_1(x,u)h_2(y,v) \qquad (7.92)$$

where

$$g_1(x,u) = h_1(x,u) = \frac{1}{\sqrt{N}} \prod_{i=0}^{n-1} (-1)^{b_i(x)b_{n-1-i}(u)} \qquad \text{and}$$

$$g_2(y,v) = h_2(y,v) = \frac{1}{\sqrt{N}} \prod_{i=0}^{n-1} (-1)^{b_i(y)b_{n-1-i}(v)}$$

That is, both the computation of a two-dimensional Walsh transform $W(u,v)$ and the computation of its inverse transform can be done by successive applications of a one-dimensional Walsh transform, and one algorithm can be used for all those computations. The procedures in computation will be the same as for Fourier transform. Analogous to the fast Fourier transform, a fast algorithm in the form of successive doubling can also be written for the Walsh transform. If the multiplying factors 1, W, W^2, ... in FFTs are all replaced by 1, the form of the fast Walsh transform and that of the fast Fourier transform will be similar.

7.5.2 Hadamard Transform

If the function

$$g(x,u) = \frac{1}{N} (-1)^{\sum_{i=0}^{n-1} b_i(x)b_i(u)} \tag{7.93}$$

is used for the forward transform kernel in the generalized transformation equation (7.10), the transformation is known as the Hadamard transform. Thus the Hadamard transform pair is

$$F_H(u) = \frac{1}{N} \sum_{x=0}^{N-1} f(x)(-1)^{\sum_{i=0}^{n-1} b_i(x)b_i(u)} \qquad u = 0, 1, \ldots, N-1 \tag{7.94}$$

$$f(x) = \sum_{u=0}^{N-1} F_H(u)(-1)^{\sum_{i=0}^{n-1} b_i(x)b_i(u)} \qquad x = 0, 1, \ldots, N-1 \tag{7.95}$$

where N is the number of samples and is also assumed to be 2^n, with n a positive integer. The same arguments on $b_k(z)$ as those used for the Walsh transform also apply to this transform.

Some properties of the H matrices are useful in their generation:

Property 1: A Hadamard matrix is a square matrix whose rows (and columns) are orthogonal with elements either +1 or −1. For an N × N matrix,

$$H_N H_N^T = NI \tag{7.96}$$

and

$$H_N = H_N^T \tag{7.97}$$

where H_N and H_N^T denote, respectively, a Hadamard matrix and its transpose, and I is an identity matrix. The lowest-order H matrix (i.e., for N = 2) is defined as

$$H_2 = \begin{vmatrix} 1 & 1 \\ 1 & -1 \end{vmatrix} \tag{7.98}$$

Property 2: $\quad H_N^{-1} = \dfrac{1}{N} H_N \tag{7.99}$

Property 3: A simple recursive algorithm can be used for the construction of the Hadamard transformation matrices, namely

$$H_{2N} = \begin{vmatrix} H_N & H_N \\ H_N & -H_N \end{vmatrix} \tag{7.100}$$

where H_N and H_{2N} represent matrices of order N and 2N, respectively If "+" and "−" are used, respectively, for the "+1" and "−1" entries for notation simplification, then

$$H_2 = \begin{vmatrix} + & + \\ + & - \end{vmatrix}$$

and

$$H_4 = \begin{vmatrix} + & + & + & + \\ + & - & + & - \\ + & + & - & - \\ + & - & - & + \end{vmatrix}$$

By the same recursive relation, we can write H_8 as

$$
H_8 = \begin{vmatrix} H_4 & H_4 \\ H_4 & -H_4 \end{vmatrix} =
\begin{vmatrix}
+ & + & + & + & + & + & + & + \\
+ & - & + & - & + & - & + & - \\
+ & + & - & - & + & + & - & - \\
+ & - & - & + & + & - & - & + \\
+ & + & + & + & - & - & - & - \\
+ & - & + & - & - & + & - & + \\
+ & + & - & - & - & - & + & + \\
+ & - & - & + & - & + & + & -
\end{vmatrix}
\begin{matrix}
\text{sequence} \\
0 \\
7 \\
3 \\
4 \\
1 \\
6 \\
2 \\
5
\end{matrix}
\tag{7.101}
$$

and H_{16} as

sequence

$$
H_{16} = \begin{vmatrix} H_8 & H_8 \\ H_8 & -H_8 \end{vmatrix} =
\begin{vmatrix}
+\ +\ +\ + & +\ +\ +\ + & +\ +\ +\ + & +\ +\ +\ + \\
+\ -\ +\ - & +\ -\ +\ - & +\ -\ +\ - & +\ -\ +\ - \\
+\ +\ -\ - & +\ +\ -\ - & +\ +\ -\ - & +\ +\ -\ - \\
+\ -\ -\ + & +\ -\ -\ + & +\ -\ -\ + & +\ -\ -\ + \\
+\ +\ +\ + & -\ -\ -\ - & +\ +\ +\ + & -\ -\ -\ - \\
+\ -\ +\ - & -\ +\ -\ + & +\ -\ +\ - & -\ +\ -\ + \\
+\ +\ -\ - & -\ -\ +\ + & +\ +\ -\ - & -\ -\ +\ + \\
+\ -\ -\ + & -\ +\ +\ - & +\ -\ -\ + & -\ +\ +\ - \\
+\ +\ +\ + & +\ +\ +\ + & -\ -\ -\ - & -\ -\ -\ - \\
+\ -\ +\ - & +\ -\ +\ - & -\ +\ -\ + & -\ +\ -\ + \\
+\ +\ -\ - & +\ +\ -\ - & -\ -\ +\ + & -\ -\ +\ + \\
+\ -\ -\ + & +\ -\ -\ + & -\ +\ +\ - & -\ +\ +\ - \\
+\ +\ +\ + & -\ -\ -\ - & -\ -\ -\ - & +\ +\ +\ + \\
+\ -\ +\ - & -\ +\ -\ + & -\ +\ -\ + & +\ -\ +\ - \\
+\ +\ -\ - & -\ -\ +\ + & -\ -\ +\ + & +\ +\ -\ - \\
+\ -\ -\ + & -\ +\ +\ - & -\ +\ +\ - & +\ -\ -\ +
\end{vmatrix}
\begin{matrix}
0 \\ 15 \\ 7 \\ 8 \\ 3 \\ 12 \\ 4 \\ 11 \\ 1 \\ 14 \\ 4 \\ 11 \\ 2 \\ 13 \\ 5 \\ 10
\end{matrix}
$$

$$(7.102)$$

The H matrix formed from the recursive construction algorithm is unordered in sequence (i.e., the number of sign changes in the rows/columns is unordered). This can be reordered by making a change in the kernel, the details of which are discussed later.

The two-dimensional Hadamard transform can be formulated as

$$F_H(u,v) = H(u,v)f(x,y)H(u,v) \tag{7.103}$$

where $F_H(u,v)$ is the Hadamard transform of $f(x,y)$ and $H(u,v)$ is the $N \times N$ symmetric Hadamard transformation matrix. The inverse Hadamard transform of $F_H(u,v)$ is

$$H(u,v)F_H(u,v)H(u,v) \tag{7.104}$$

or

$$H(u,v)H(u,v)f(x,y)H(u,v)H(u,v)$$

after substitution of $F_H(u,v)$ from Eq. (7.103). By using the relation expressed on Eq. (7.96), we have

$$H(u,v)F_H(u,v)H(u,v) = N^2 f(x,y) \tag{7.105}$$

Thus

$$f(x,y) = \frac{1}{N^2} H(u,v)F_H(u,v)H(u,v) \qquad (7.106)$$

which forms a Hadamard transformation pair with Eq. (7.103).

In order to put the sequence in increasing order, let us let the forward transformation kernel be of the following form:

$$g(x,y;u,v) = \frac{1}{N} (-1)^{\sum_{i=0}^{n-1}[b_i(x)p_i(u)+b_i(y)p_i(v)]} \qquad (7.107)$$

The Hadamare transform becomes

$$F_H(u,v) = \frac{1}{N} \sum_{x=0}^{N-1} \sum_{y=0}^{N-1} f(x,y)(-1)^{\sum_{i=0}^{n-1}[b_i(x)p_i(u)+b_i(y)p_i(v)]} \qquad (7.108)$$

where both x and u are in binary representation. $b_k(z)$ represents the kth bit in the binary representation of z with the zeroth bit as the least significant one. $p_\ell(r)$ is defined as follows:

$$p_0(u) = b_{n-1}(u)$$

$$p_1(u) = b_{n-1}(u) + b_{n-2}(u)$$

$$p_2(u) = b_{n-2}(u) + b_{n-3}(u) \qquad (7.109)$$

$$\vdots$$

$$p_{n-1}(u) = b_1(u) + b_0(u)$$

and

$$\sum_{i=0}^{n-1} p_i(u) = b_0(u) \quad (\text{mod } 2) \qquad (7.110)$$

where $b_{n-1}(u)$ represents the leftmost binary bit of u; $b_{n-2}(u)$, the next-leftmost bit of u; and so on. The summations in Eqs. (7.109) and (7.110) are performed in modulo 2 arithmetic. Similar arguments apply to $p_i(v)$, i = 0, 1, ..., n − 1.

Example For the one-dimensional Hadamard transform, compute the values of the ordered Hadamard kernel for $N = 8$ (or $n = 3$).

Solution: When $u = 2$ (0 1 0 in binary representation) and $x = 6$ (1 1 0 in binary), we have

$$p_0(u) = 0$$

$$p_1(u) = b_2(u) + b_1(u) = 0 + 1 = 1$$

$$p_2(u) = b_1(u) + b_0(u) = 1 + 0 = 1$$

$$\sum_{i=0}^{n-1} b_i(x)p_i(u) = b_0(x)p_0(u) + b_1(x)p_1(u) + b_2(x)p_2(u)$$

$$= 0 + 1 + 1$$

$$= 2$$

when $u = 5$ (1 0 1 in binary representation) and $x = 4$ (1 0 0), we have

$$p_0(u) = 1$$

$$p_1(u) = b_2(u) + b_1(u) = 1 + 0 = 1$$

$$p_2(u) = b_1(u) + b_0(u) = 0 + 1 = 1$$

$$\sum_{i=0}^{n-1} b_i(x)p_i(u) = b_0(x)p_0(u) + b_1(x)p_1(u) + b_2(x)p_2(u)$$

$$= 0 + 0 + 1 = 1$$

and so forth, and an ordered Hadamard transformation kernel can be constructed as shown in (7.111).

u \ x	0	1	2	3	4	5	6	7	sequence
0	+	+	+	+	+	+	+	+	0
1	+	+	+	+	−	−	−	−	1
2	+	+	−	−	−	−	+	+	2
3	+	+	−	−	+	+	−	−	3
4	+	−	−	+	+	−	−	+	4
5	+	−	−	+	−	+	+	−	5
6	+	−	+	−	−	+	−	+	6
7	+	−	+	−	+	−	+	−	7

(7.111)

By comparing (7.111) with (7.101), we can see that the sequence in (7.111) is ordered.

As can be seen from Eq. (7.107), the kernel for the two-dimensional ordered Hadamard transform is separable. Thus

$$F_H(u,v) = \frac{1}{N} \sum_{y=0}^{N-1} \left[\sum_{x=0}^{N-1} f(x,y)(-1)^{\sum_{i=0}^{n-1} b_i(x)p_i(u)} (-1)^{\sum_{i=0}^{n-1} b_i(y)p_i(v)} \right]$$

$$= \frac{1}{\sqrt{N}} \sum_{y=0}^{N-1} F_H(u,y)(-1)^{\sum_{i=0}^{n-1} b_i(y)p_i(v)} \qquad (7.112)$$

where $F_H(u,y)$ is a one-dimensional Hadamard transform. Analogous to the two-dimensional FFT, the one-dimensional Hadamard transform can be successively used for the two-dimensional transformation and a fast algorithm can also be established for it. Analogous to the Fourier transform, the Hadamard transform as expressed by

$$F_H(u) = \sum_{x=0}^{N-1} f(x)(-1)^{\sum_{i=0}^{n-1} b_i(x)p_i(u)} \qquad (7.113)$$

can be decomposed into the sum of two series of terms as

$$F_H(u) = G(u) + (-1)^{\sum_{i=0}^{n-1} p_i(u)} H(u) \qquad (7.114)$$

where

$$G(u) = \sum_{r=0}^{\frac{N}{2}-1} f(2r)(-1)^{\sum_{i=0}^{n-1} b_i(2r)p_i(u)} \qquad (7.115)$$

and

$$H(u) = \sum_{r=0}^{\frac{N}{2}-1} f(2r+1)(-1)^{\sum_{i=0}^{n-1} b_i(2r)p_i(u)} \qquad (7.116)$$

7.5.3 Discrete Karhunen–Loéve Transform

Detailed discussion of this transform has already been given in Sec. 6.3. What we are going to add here is the application of this transform in image processing. Let us put the $N \times N$ matrix $f(x,y)$,

$$f(x,y) = \begin{vmatrix} f(0,0) & f(0,1) & \cdots & f(0,\,N-1) \\ f(1,0) & & & \\ \cdot & & & \\ \cdot & & & \\ f(N-1,\,0) & \cdots & & f(N-1,\,N-1) \end{vmatrix} \qquad (7.117)$$

into the form of an N^2-element vector as expressed by

$$\underline{x}_i = \begin{vmatrix} x_{i1} \\ x_{i2} \\ \cdot \\ \cdot \\ \cdot \\ x_{ij} \\ \cdot \\ \cdot \\ \cdot \\ x_{iN^2} \end{vmatrix} \qquad (7.118)$$

where $\underline{x}_i$, $i = 1, 2, \ldots, K$, represent image samples, and x_{i1}, x_{i2}, $\ldots$, x_{iN^2} correspond respectively to $f(0, 0)$, $f(0, 1)$, $\ldots$, $f(N - 1, N - 1)$ of the ith image sample. The transform can then be treated as a statistical problem. Following the discussions in Chap. 4, we have

$$\underline{C}_x = E\{(\underline{x} - \underline{m}_x)(\underline{x} - \underline{m}_x)^T\} \qquad (7.119)$$

where $\underline{C}_x$ is the covariance matrix, and $\underline{m}_x$ the mean value of $\underline{x}$, both of which can be approximated by

$$\underline{m}_x \simeq \frac{1}{K} \sum_{i=1}^{k} \underline{x}_i \qquad (7.120)$$

and

$$\underline{C}_x \simeq \frac{1}{K} \sum_{i=1}^{K} \underline{x}_i \underline{x}_i^T - \underline{m}_x \underline{m}_x^T \qquad (7.121)$$

where K is the number of image samples, $\underline{m}_x$ is an N^2 vector, and $\underline{C}_x$ is an $N^2 \times N^2$ matrix. The problem we now have is to transform the original image vector $\underline{x}$ into a new image vector $\underline{y}$ so that the covariance matrix $\underline{C}_y$ will be a diagonal one. Thus we have

$$\underline{y} = \underline{B}(\underline{x} - \underline{m}_x) \tag{7.122}$$

where $(\underline{x} - \underline{m}_x)$ represents the centralized image vector, and the $N^2 \times N^2$ matrix $\underline{B}$ is chosen such that its rows are eigenvectors of $\underline{C}_x$; thus

$$\underline{B} = \begin{vmatrix} \underline{e}_1 \\ \underline{e}_2 \\ \cdot \\ \cdot \\ \cdot \\ \underline{e}_i \\ \cdot \\ \cdot \\ \cdot \\ \underline{e}_{N^2} \end{vmatrix} = \begin{vmatrix} e_{11} & & \cdots & e_{1N^2} \\ e_{21} & & \cdots & e_{2N^2} \\ \cdot & & & \cdot \\ \cdot & & & \cdot \\ \cdot & & & \cdot \\ e_{i1} & & & e_{iN^2} \\ \cdot & & & \cdot \\ \cdot & & & \cdot \\ \cdot & & & \cdot \\ e_{N^2 1} & e_{N^2 2} & \cdots & e_{N^2 N^2} \end{vmatrix} \tag{7.123}$$

where $\underline{e}_i = [e_{i1}, \ldots, e_{iN^2}]$ is the ith eigenvector of $\underline{C}_x$ and e_{ij} is the jth component of the ith eigenvector. The new covariance matrix $\underline{C}_y$ is then

$$\underline{C}_y = E\{B(\underline{x} - \underline{m}_x)(\underline{x} - \underline{m}_x)^T \underline{B}^T\}$$

$$= \underline{B}\underline{C}_x\underline{B}^T \tag{7.124}$$

which is a diagonal matrix, for the reasons given below. Since

$$\underline{y} = \underline{B}(\underline{x} - \underline{m}_x) \tag{7.125}$$

then

$$\underline{x} - \underline{m}_x = \underline{B}^T\underline{y} \tag{7.126}$$

where $\underline{y} = [y_1, y_2, \ldots, y_p]$ and $\underline{B}$ is an orthogonal matrix. Let $\underline{B}_r$ denote the rth column of $\underline{B}$ (and $\underline{B}_r$ the rth row of $\underline{B}^T$). Then $\underline{B}_1$ is chosen first in such a way that the variance of y_1 is maximized; $\underline{B}_2$ is chosen so that the variance of y_2 is maximized subject to the condition that y_2 is uncorrelated with y_1; and similarly for the remaining y's. The variance of y_r is maximized subject to the condition that y_r is uncorrelated with $y_1, y_2, \ldots, y_{r-1}$. Let us denote the variance of y_r by λ_r. Since $\underline{y}_r = \underline{B}_r^T \underline{x}$, we have

$$\lambda_r = \underline{B}_r^T \underline{C}_x \underline{B}_r \tag{7.127}$$

As the y's are uncorrelated, we also have

$$\underline{B}_r^T \underline{C}_x \underline{B}_s = 0 \quad \text{for } r \neq s \tag{7.128}$$

This means that

$$\underline{B}^T \underline{C}_x \underline{T} = \Lambda \tag{7.129}$$

which is diagonal with elements $\lambda_1, \lambda_2, \ldots, \lambda_p$ arranged in order of magnitude. So we have

$$
\underline{C}_y =
\begin{vmatrix}
\lambda_1 & & & & & & \\
 & \lambda_2 & & & & 0 & \\
 & & \ddots & & & & \\
 & & & \ddots & & & \\
 & & & & \lambda_i & & \\
 & 0 & & & & \ddots & \\
 & & & & & & \lambda_{N^2}
\end{vmatrix}
\tag{7.130}
$$

with elements equal to the eigenvalues of $\underline{C}_x$ (λ_i, $i = 1, 2, \ldots, N^2$), where λ_i is the variance of the ith element of $\underline{y}$ along eigenvector $\underline{e}_i$. $\underline{x}$ can be reconstructed from $\underline{y}$ by using

$$\underline{x} = \underline{B}^T \underline{y} + \underline{m}_x \tag{7.131}$$

This is because $\underline{B}^{-1} = \underline{B}^T$ for orthonormal vectors.

The Karhunen-Loéve transform is useful in data compression and image rotation applications. But this transform has the drawback of not being separable, and therefore no fast algorithm exists for computing the transform.

PROBLEMS

7.1 Show that the Fourier transform of a rectangular function with magnitude A is a sinc function. Sketch the resulting Fourier spectrum.

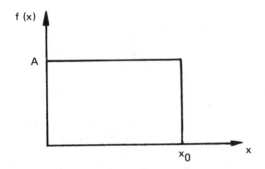

7.2 Note that in Fig. 7.17, the f's and F's are ordered differently in order to retain the butterfly computation format. Work out the flow graph of the FFT algorithm with the inputs in natural order.

7.3 Derive an equivalent algorithm for the FFT decimation-in-frequency decomposition of an eight-point DFT computation.

7.4 (a) Construct the unordered H matrix for N = 16 and mark the number of sign changes in each row. (b) Rearrange the H matrix in part (a) so that the sequence is in increasing order.

8
Image Enhancement

Image enhancement is one of the important steps in the processing of large data sets to make the results more suitable for classification than were the original data. Although lots of methods have been suggested for image enhancement, there are still no general approaches available that can be followed for every case. The reason is that the enhancement of an image is very problem oriented. Near the end of this chapter, an approach along these lines is suggested, and the results obtained have been encouraging, but more work must be done before the method becomes mature.

The approaches suggested for enhancement can be grouped into two categories: spatial processing and transform processing. In transform processing, the image function is first transformed to the transform domain and then processed to meet specific problem requirements. Inverse transform is needed to yield the final spatial image results. On the other hand, with spatial domain processing, the pixels in the image are manipulated directly.

8.1 ENHANCEMENT BY SPATIAL PROCESSING

Processing in the spatial domain is usually carried out pixel by pixel. Depending on whether the processing is based only on the processed pixel itself or is taking its neighboring pixels into consideration while processing, the processing can be divided into two subcategories: point processing and neighborhood processing. In neighborhood processing, 3 × 3 or 5 × 5 windows are frequently used for the processing of a single pixel, and it is quite obvious that the computational time will be greatly increased. Nevertheless, an increase in computational time is sometimes required to obtain local context information, which is useful for decision making in specific pixel processing for certain purposes.

8.1.1 Point Processing in the Spatial Domain

Deterministic Gray-Level Transformation

By *point processing* we mean that the processing of a certain pixel in the image depends on the information we have on that pixel itself without consideration of the status of its neighborhood. There are several ways to treat this problem, one of which is deterministic gray-level transformation, and the other, histogram modification. Deterministic gray-level transformation is quite straightforward. A conversion table or algebraic expression will be stored, and the gray-level transformation for each pixel will be carried out either by the table lookup or by algebraic computation, as shown schematically in Fig. 8.1., in which $g(x,y)$ is the image after gray-level transformation of the original image $f(x,y)$. For an image of 512×512 pixels, 262,144 operations will be required. Figure 8.2 shows some of the deterministic gray-level transformations that could be used to meet various requirements. With the gray-level transformation function as shown in part (a), the straight-line function "1" yields a brighter output than the original, whereas the straight-line function "2' gives a lower gray-level output for each pixel of the original image. Part (b) shows brightness stretching on the midregion of the image gray levels. Part (c) gives a more accentric action on this transformation which is useful in contrast stretching, while part (d) is the limiting transformation function, which yields a binary image (i.e., only two gray levels would exist in the image). Part (e) gives the opposite effect to that shown in part (b). In part (f), function "1" shows a dark region stretching transformation (i.e., dark becomes less dark, bright becomes less bright); while function "2" gives a bright region stretching transformation (i.e., lower gray-level outputs for lower gray levels of f, but higher gray-level outputs for higher gray levels of f). The sawtooth contrast scaling gray-level transformation function shown in part (g) can be used to produce a wide-dynamic-range image on a small-dynamic-range display. This is achieved by removing the most significant bit of the pixel value. Part (h) shows a reverse scaling, by means of which a negative of the original image can be obtained. Part (i) shows a

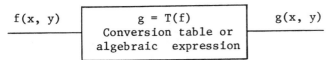

FIG. 8.1 Schematic diagram of the deterministic gray-level transformation.

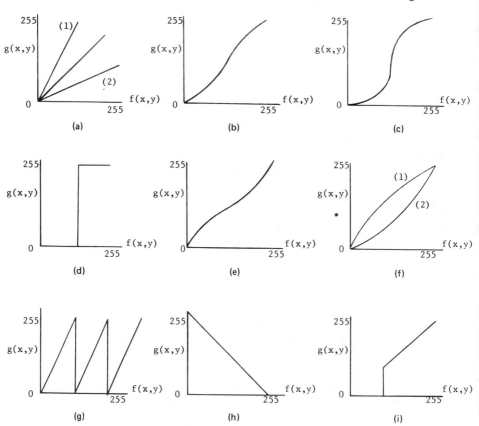

FIG. 8.2 Various gray-level transformation functions.

thresholding transformation, where the height h can be changed to adjust the output dynamic range.

Histogram Modification

Histogram modification is another effective method used in point processing. By *histogram*, we mean the plot of the probability density (or the relative occurrence frequency) $P_r(r)$ versus r, where $r = f(x,y)$ normalized within the range $(0,1)$. Figure 8.3(a) shows a histogram plot and Fig. 8.3b shows the cumulative density function versus r plot. With the histogram a distribution of gray levels of pixels in an image can be described.

For an image of $N \times N$ pixels, the total number of pixels in the image is $\sum_{i=r_1}^{r_k} n_i = N^2$, where $r_1, \ldots, r_k$ are the gray levels and n_i

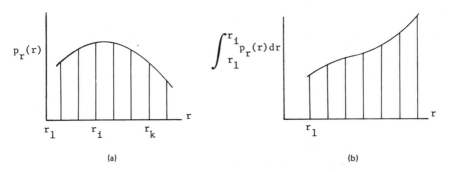

(a)   (b)

FIG. 8.3 Probability density function and cumulative density function versus r plot: (a) $p_r(r)$ versus r plot (histogram); (b) cumulative probability density function versus r plot.

is the number of pixels at gray level r_i. The histogram and the cumulative probability density function (CPDF) will be of the form shown in Fig. 8.4a and b, respectively.

It is obvious that $\sum_{r_1}^{r_k}$ (PDF) = 1, and CPDF is a single-valued monotonic function. If the input image intensity variable $r = f(x, y)$, $r_0 \leqslant r \leqslant r_J$, for the original image is mapped into the output image intensity $s = g(x,y)$, $s_0 \leqslant s \leqslant s_k$, for the processed image such that the output probability distribution $P_s(s_k)$ follows some desired form for a given input probability distribution $P_r(r_j)$, we can relate them by Eqs. (8.1) and (8.2):

$$s = T(r) \quad r_0 \leqslant r \leqslant r_J \tag{8.1}$$

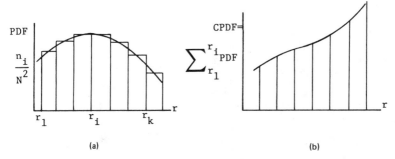

(a)   (b)

FIG. 8.4 Probability density function and the cumulative probability density function versus r plot: (a) histogram; (b) cumulative probability density function versus r plot.

or

$$r = T^{-1}(s) \quad s_0 \leqslant s \leqslant s_K \tag{8.2}$$

This histogram modification problem can then be formulated as follows:

1. Find the transformation function $T(r)$ to relate the PDF of the image on the original gray-level scale and the desired probability density distribution of the image on a new gray-level scale; or
2. Find the probability density function PDF of the image on a new gray-level scale, when the PDF of the image on the original gray-level scale and the transformation function are given.

Note that this transformation is a gray-level transformation only. The number of gray levels may be different before and after the transformation, but there is no loss in pixel numbers. Hence

$$\sum_{j=0}^{J} p_r(r_j) = 1 \tag{8.3}$$

$$\sum_{k=0}^{K} p_s(s_k) = 1 \tag{8.4}$$

Equations (8.3) and (8.4) state that the input and output probability distributions must both sum to unity. Furthermore, the cumulative distributions for the input and output must equate for any input index j, that is,

$$\sum_{k} p_s(s_k) = \sum_{j} p_r(r_j) \tag{8.5}$$

Thus the probability that pixels in the input image have a pixel luminance value $\leqslant r_j$ must be equal to the probability that pixels in the output image have a pixel luminance value $\leqslant s_k$, as long as the transformation rule $s_k = T(r_j)$ is followed. The transformation in this case is monotonic. If for a given image the cumulative probability distribution $\Sigma_j \, p_r(r_j)$ is replaced by the cumulative histogram $\Sigma H_r(j)$, a solution for s_k in terms of r_j can be obtained by an inverted form of Eq. (8.5), as shown in Fig. 8.5.

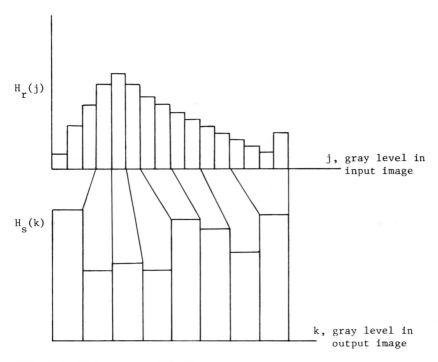

FIG. 8.5 Histogram modification.

Although it is not impossible, it is very difficult to solve this problem analytically except for the very simple uniform output histogram case. Let us replace the summation by integration; we then have

$$\int_{S_{min}}^{S} p_s(s)\ ds = \int_{r_{min}}^{r} p_r(r)\ dr \tag{8.6}$$

where the definite integral of $p_r(r)$ over r on the right-hand side is the cumulative distribution of the input image and equal, say, to $P_r(r)$.

For a uniform output histogram,

$$p_s(s) = const. = \frac{1}{s_{max} - s_{min}} \quad \text{for}\ s_{min} \leqslant s \leqslant s_{max} \tag{8.7}$$

Substitution of Eq. (8.7) in (8.6) gives

$$\int_{S_{min}}^{S} \frac{1}{s_{max} - s_{min}} \, ds = \int_{r_{min}}^{r} p_r(r) \, dr \tag{8.8}$$

or

$$\frac{s - s_{min}}{s_{max} - s_{min}} = P_r(r) \tag{8.9}$$

from which we have

$$s = [s_{max} - s_{min}]P_r(r) + s_{min} \tag{8.10}$$

which is the histogram equalization transfer function and is shown in Fig. 8.6. $P_r(r)$, the cumulative probability distribution of the input image, can be approximated by its cumulative histogram or $P_r(r) \simeq \sum_j H_r(j)$. For an exponential output histogram,

$$p_s(s) = \alpha e^{-\alpha[s-s_{min}]} \qquad s \geqslant s_{min} \tag{8.11}$$

By substituting Eq. (8.11) in Eq. (8.6), we obtain

$$\int_{S_{min}}^{S} \alpha e^{-\alpha[s-s_{min}]} \, ds = \int_{r_{min}}^{r} p_r(r) \, dr \tag{8.12}$$

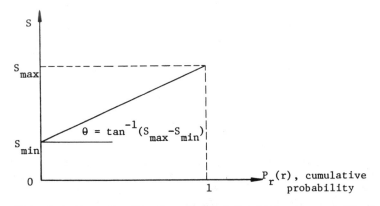

FIG. 8.6 Transfer function derived for histogram equalization.

or

$$1 - e^{-\alpha[s-s_{min}]} = P_r(r) \tag{8.13}$$

The transfer function will then be

$$s = s_{min} - \frac{1}{\alpha} \ln[1 - P_r(r)] \tag{8.14}$$

The procedure of conducting the histogram equalization can be summarized as consisting of the following steps:

1. Compute the average number of pixels per gray level.
2. Starting from the lowest gray-level band, accumulate the number of pixels until the sum is closest to the average. All

TABLE 8.1 Example Image Showing Histrogram Equalization Mapping

k	r_k	n_k	$p_r(r_k) = n_k/\Sigma n_k$	CPDF $= \Sigma P_r(r_k)$
0	0	300	0.0183	0.0183
1	1/15	1,500	0.0916	0.1099
2	2/15	3,500	0.2136	0.3235
3	3/15	3,000	0.1831	0.5066
4	4/15	2,125	0.1297	0.6363
5	5/15	1,625	0.0992	0.7355
6	6/15	1,250	0.0763	0.8118
7	7/15	900	0.0549	0.8667
8	8/15	650	0.0397	0.9064
9	9/15	550	0.0336	0.9400
10	10/15	325	0.0198	0.9598
11	11/15	200	0.0122	0.9720
12	12/15	150	0.0092	0.9812
13	13/15	140	0.0085	0.9897
14	14/15	97	0.0059	0.9956
15	1	72	0.0044	1.0000
		$\Sigma n_k = 16,384$	1.0000	

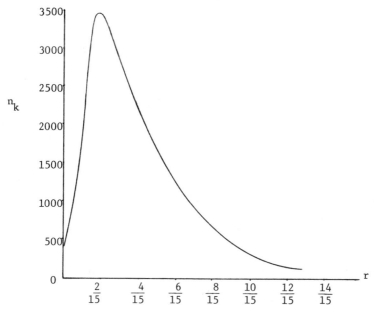

FIG. 8.7 Original histogram of the example image.

of these pixels are then rescaled to the new reconstruction
levels.

3. If an old gray-level band is to be divided into several new
bands, either do it randomly or adopt a rational strategy—
one being to distribute them by region. An example of a 16-
gray-level 128 × 128 pixel image is shown in Table 8.1.
Sixteen equally spaced gray levels are assumed in this example. The average number of pixels per gray level is
16,384/16 = 1024. The histogram of this image is shown in
Fig. 8.7, and the gray-level transformation matrix can be
formulated as shown in Fig. 8.8.

8.1.2 Neighborhood Processing in the Spatial Domain

Neighborhood processing differs from Point processing in that the local
contextual information will be used for specific pixel processing. Pixels
close to each other are supposed to have approximately the same gray
levels except for those at the boundary.

Smoothing and Noise Elimination

Noise can come from different sources. It can be introduced during
transmission through the channel. This kind of noise has no relation

New gray levels

Original gray levels	0	1/15	2/15	3/15	4/15	5/15	6/15	7/15	8/15	9/15	10/15	11/15	12/15	13/15	14/15	15/15	Summation of pixels
0	300																300
1/15	724	776															1500
2/15		248	1024	1024	1024	180											3500
3/15						844	1024	1024	108								3000
4/15									916	1024	185						2125
5/15											839	786					1625
6/15												238	1012				1250
7/15													12	888			900
8/15														136	514		650
9/15															510	40	550
10/15																325	325
11/15																200	200
12/15																150	150
13/15																140	140
14/15																97	97
1																72	72
	1024	1024	1024	1024	1024	1024	1024	1024	1024	1024	1024	1024	1024	1024	1024	1024	16384

FIG. 8.8 Gray-level transformation matrix for the example image of Fig. 8.7.

to the image signal. Its value is generally independent of the strength of the image signal. It is additive in nature and can be put in the following form:

$$f(x,y) = f'(x,y) + n(x,y)$$

where $f'(x,y)$ simply denotes the hypothetically noise-free image; n, the noise; and $f(x,y)$, the noisy image. This kind of noise can be estimated by a statistical analysis of a region known to contain a constant gray level, and can be minimized by classical statistical filtering or by spatial ad hoc processing technique.

Another source of noise is the equipment or the recording medium. This kind of noise is multiplicative in nature, and its level depends on the level of the image signal. An example of this is noise from the digitizer, such as the flying spot scanner, TV raster lines, and photographic grain noise.

In practice, there frequently is a difference between an image and its quantized image. This difference can be classified as quantization noise and can be estimated.

In the smoothing of an image, we have to determine the nature of the noise points first. If they belong to fine noise, they are usually isolated points. This means that each noise point has nonnoise neighbors. These noise points, known as "salt-and-pepper noise," usually occur on raster lines.

After determining the nature of the noise, we can set up an approximate approach to eliminate the noise. Detection of the noise point can be done by comparing its gray level with those of its neighbors. If its gray level is substantially larger than or smaller than those of all or nearly all of its neighbors, the point can be classified as a noise point, and the gray level of this noise point can be replaced by the weighted average of the gray levels of its neighbors.

Coarse noise (e.g., a 2 × 2 pixel core) is rather difficult to deal with. The difficulty will be in detection. We treat such noise by various appropriate methods, and sometimes we need some a priori information.

Various sizes of neighborhood can be used for noise detection and for the smoothing process. A 3 × 3 pixel window is satisfactory in most cases, although a large neighborhood is sometimes used for statistical smoothing. For the border points (i.e., for the points on four sides of an image), some amendment should be added either by ignoring or by duplicating the side rows and columns.

Multiples of the estimated standard deviation of the noise can be chosen for the threshold by which the noisy point must differ from its neighborhood. The estimation of the standard deviation of the noise can be obtained by measuring the standard deviation of gray levels over a region that is constant in the nonnoisy image. It is assumed here that the noise has zero mean. Some other method, such as a majority count of the neighbors that have larger or smaller gray levels than the given point, is also adopted.

Based on the arguments we have presented so far, a decision rule can be set for the determination of the gray level of each point. Generally speaking, if a point is a noise point, its gray level can be replaced by the weighted average of its neighbors. If a point is not a noise point, its original gray level is still used.

There are some other possible ways of deciding whether the given point is a noise point:

1. Compare the gray level of the point with those of its neighbors. If the comparison shows that $|f - f_i| \geqslant \tau$, it will be considered a noise point, where f is the gray level of the point; f_i, $i = 0$, ..., 8, if a 3 × 3 neighborhood is used, is the gray levels of its neighboring points; and τ is the threshold chosen.

2. Compare the gray level of the point with the average gray level of its neighbors. If $|f - f_{avg}| \geqslant 0$, the point is a noise point. The advantage of this algorithm is simplicity of computation; its shortcoming is that some difficulties will be experienced in distinguishing isolating points from points on edges or on boundary lines.

3. A third way of determining whether the point is a noise point is called "fuzzy decision." Let p be the probability of the point being a noise point; then $(1 - p)$ will be the probability of the point not being a noise point. Then a linear combination of f, the gray level of that point, and f_{avg}, the average gray level of its neighbors ($f_{avg} = \Sigma\, f_i/k$) will give the gray level of the point under consideration, such as

$$g = (1 - p)f + p\, \frac{\Sigma\, f_i}{k} \qquad (8.15)$$

A straightforward neighborhood averaging algorithm is discussed here for the smoothing processing. Let f(x,y) be the noisy image. The smoothed image g(x,y) after averaging processing can be obtained from the following relation:

$$g(x,y) = \frac{1}{N} \sum_{n,m \in \Omega} \sum W_{n,m} f(n,m) \qquad \text{for x, y = 0, 1, ..., N - 1}$$

$$(8.16)$$

where Ω is the rectangular n × m neighborhood defined, $W_{n,m}$ the n × m weight array, and N the total number of pixels defined by Ω. A criterion can be formulated such that if

$$|f(x,y) - g(x,y)| \geqslant \tau \qquad (8.17)$$

where τ denotes the threshold chosen, then $f(x,y)$ is replaced by $g(x,y)$. Otherwise, no change in the gray level will be made.

If, for example, the weight array $W_{n,m}$ and $f(x,y)$ are, respectively,

$$\frac{1}{N} W_{n,m} = \frac{1}{9} \begin{vmatrix} 1 & 1 & 1 \\ 1 & 1 & 1 \\ 1 & 1 & 1 \end{vmatrix}$$

and

$$f(x,y) = \begin{vmatrix} 100 & 120 & 100 \\ 120 & (200) & 90 \\ 100 & 90 & 100 \end{vmatrix}$$

and the threshold value chosen for the example is 40, then

$$g(x,y) = \begin{vmatrix} 100 & 120 & 100 \\ 120 & (113) & 90 \\ 100 & 90 & 100 \end{vmatrix}$$

where the circled pixel "200" is replaced by the average gray level "113" of the nine pixels. Note that the multiplication $\Sigma\Sigma \, W_{n,m}f(n,m)$ in Eq. (8.16) is not a matrix multiplication, but is

$$g(x,y) = \frac{1}{N} \sum_{n,m=1}^{3}\sum W_{n,m}f(n,m)$$

$$= \frac{1}{9} \begin{vmatrix} W_{11}f(1,1) & W_{12}f(1,2) & W_{13}f(1,3) \\ W_{21}f(2,1) & W_{22}f(2,2) & W_{23}f(2,3) \\ W_{31}f(3,1) & W_{32}f(3,2) & W_{33}f(3,3) \end{vmatrix} \qquad (8.18)$$

where

$$W_{n,m} = \begin{vmatrix} W_{11} & W_{12} & W_{13} \\ W_{21} & W_{22} & W_{23} \\ W_{31} & W_{32} & W_{33} \end{vmatrix}$$

and

$$f(n,m) = \begin{vmatrix} f(1,1) & f(1,2) & f(1,3) \\ f(2,1) & f(2,2) & f(2,3) \\ f(3,1) & f(3,2) & f(3,3) \end{vmatrix}$$

Let us take another example for illustration. Let

$$f(x,y) = \begin{vmatrix} 100 & 120 & 100 \\ 120 & 140 & 90 \\ 100 & 90 & 100 \end{vmatrix}$$

where the gray level of the pixel to be processed is 140, while those of its neighbors are the same as in the previous example. Since the absolute value of the difference between 140 and the average of the nine pixels (107) is less than the threshold, no change in gray-level value is made on the circled pixel. Move the mask over every pixel in the image, and a smoothed image can be obtained. The reason for using the threshold in the process is to reduce the blurring effect produced by the neighborhood averaging.

Some other masks were also suggested. They are shown in Fig. 8.9. Note that in parts (a) and (b), different weights are given to the centering pixel and its neighbors. In parts (c) and (d) the averaging process is operated only on the neighboring pixels and different neghbors are taken into consideration. This smoothing process can be generalized as

$$W = \frac{1}{10} \begin{vmatrix} 1 & 1 & 1 \\ 1 & 2 & 1 \\ 1 & 1 & 1 \end{vmatrix} \qquad W = \frac{1}{16} \begin{vmatrix} 1 & 2 & 1 \\ 2 & 4 & 2 \\ 1 & 2 & 1 \end{vmatrix}$$

(a)　　　　　　　　　　　　　(b)

$$W = \frac{1}{4} \begin{vmatrix} 0 & 1 & 0 \\ 1 & 0 & 1 \\ 0 & 1 & 0 \end{vmatrix} \qquad W = \frac{1}{8} \begin{vmatrix} 1 & 1 & 1 \\ 1 & 0 & 1 \\ 1 & 1 & 1 \end{vmatrix}$$

(c)　　　　　　　　　　　　　(d)

FIG. 8.9 Various masks frequently used in image smoothing.

$$g(x,y) = \begin{cases} \dfrac{1}{N} \displaystyle\sum_{n,m\in\Omega} W_{n,m} f(n,m) & \text{if } \left| f(x,y) - \dfrac{1}{N} \displaystyle\sum_{n,m\in\Omega} W_{n,m} f(n,m) \right| > \tau \\[4mm] f(x,y) & \text{otherwise} \end{cases}$$

$$(8.19)$$

Edge Sharpening

As just discussed, the smoothing of an image by neighborhood averaging is analogous to an integration process. In the integration process, part of the edge information is lost and blurring results. Differentiation is the reverse of integration, and therefore by using differentiation, a sharpening effect can be expected on the edge.

If we are given a two-dimensional image function $f(x,y)$, a vector gradient $\underline{G}[(f(x,y)]$ can be formed as

$$\underline{G}[f(x,y)] = \begin{vmatrix} \dfrac{\partial f}{\partial x} \\[3mm] \dfrac{\partial f}{\partial y} \end{vmatrix}$$

$$(8.20)$$

The direction of this vector is toward the maximum rate of increase of the image function $f(x,y)$, while its magnitude is represented by

$$\underline{G} = \left[\left(\frac{\partial f}{\partial x} \right)^2 + \left(\frac{\partial f}{\partial y} \right)^2 \right]^{1/2}$$

$$(8.21)$$

For discrete images, the coordinate system is chosen as shown in Fig. 8.10, with $\partial f / \partial x$ points in the horizontal rightward direction and $\partial f / \partial y$ points in the vertical downward direction. Equation (8.21) can then be implemented digitally by

$$\underline{G} = [(f(j,k) - f(j+1, k))^2 + (f(j,k) - f(j, k+1))^2]^{1/2}$$

$$(8.22)$$

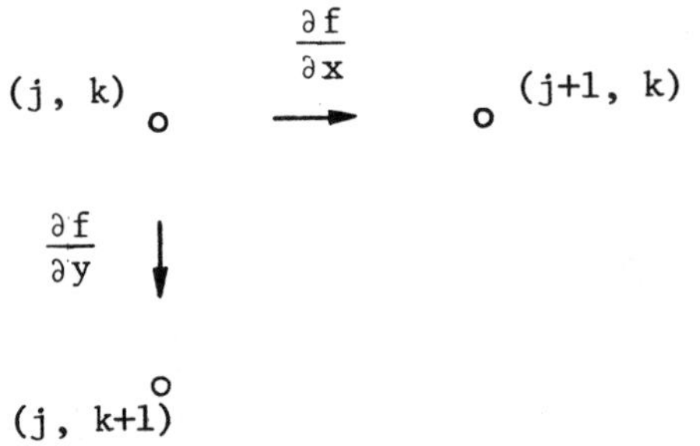

FIG. 8.10 Digital implementation of a gradient operator.

(j, k)   (j+1, k)

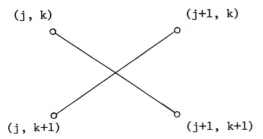

(j, k+1)   (j+1, k+1)

FIG. 8.11 Robert's cross-gradient operator.

which is commonly called a *three-point gradient*. This implementation is more accurate, but is computationally expensive. If the absolute values of the terms inside the brackets under the square root are taken for the value of $|G|$,

$$|G| \simeq |f(j,k) - f(j + 1, k)| + |f(j,k) - f(j, k + 1)| \qquad (8.23)$$

computational advantages can be achieved.

Another digital implementation for the gradient is called *Roberts' cross operator*. This is shown in Fig. 8.11, where cross differences are used for the implementation as follows:

$$|G| \simeq |f(j,k) - f(j + 1, k + 1)| + |f(j, k + 1) - f(j + 1, k)|$$

$$(8.24)$$

This is commonly called a *four-point gradient*.

It can easily be seen from Eqs. (8.22) to (8.24) that the magnitude of the gradient is large for prominent edges, small for a rather smooth area, and zero for a constant gray-level area.

Several algorithms are available for performing the edge sharpening:

Algorithm 1 Edge sharpening by selectively replacing a pixel point by its gradient. The algorithm can be put in the form shown in Fig. 8.12. The computation of $|G[f(j,k)]|$ can be done using the three-point gradient or the four-point gradient expression. When the value of $|G[f(j,k)]|$ is greater than or equal to a threshold, replace the pixel by its gradient value, or by gray level 255. Otherwise, keep it in the original level or a zero (or low) gray level. The process will run point by point through the 262,144 points from top to bottom and from left to right on the image.

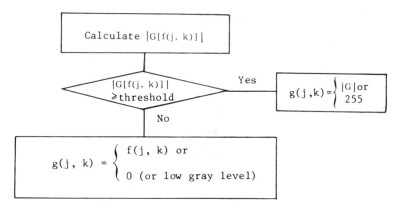

FIG. 8.12 Algorithm for edge sharpening by selectively replacing pixel points by their gradients.

Different gradient images can be generated from this algorithm depending on the values chosen for the $g(j,k)$, $j,k = 1, 2, \ldots, N - 1$, when its gradient $G[f(j,k)] \geq T$ and the values chosen for $g(j,k)$ when $G[f(j,k)] < T$. A binary gradient image will be obtained when the edges and background are displayed in 255 and 0. An edge-emphasized image without destroying the characteristics of smooth background can be obtained by

$$g(j,k) = \begin{cases} |G[f(j,k)]| & \text{if } |G[f(j,k)]| \geq T \\ f(j,k) & \text{otherwise} \end{cases}$$

Algorithm 2 Edge sharpening by statistical differencing. The algorithm can be stated as follows:

(a) Specify a window (typically 7×7 pixels) with the pixel (j,k) as the center.

(b) Compute the standard deviation σ or variance σ^2 over the elements inside the window.

$$\sigma^2(j,k) = \frac{1}{N} \sum_{j,k \in \text{window}}^{\sqrt{N}} \sum^{\sqrt{N}} [f(j,k) - \bar{f}(j,k)]^2 \tag{8.27}$$

where N is the number of pixels in the window and $\bar{f}(j,k)$ is the mean value of the original image in the window.

(c) Replace each pixel by a value $g(j,k)$, which is the quotient of its original value divided by $\sigma(j,k)$ to generate the new image, such as

$$g(j,k) = \frac{f(j,k)}{\sigma(j,k)}$$

It can be noted that the enhanced image $g(j,k)$ will be increased in amplitude with respect to the original image $f(j,k)$ at the edge point and decreased elsewhere.

Algorithm 3 Edge sharpening by spatial convolution with a high pass mask H such as

$$g(j,k) = \sum_{n} \sum_{m} f(n,m)H(n,m,j,k) \qquad (8.25)$$

Examples of a high pass mask for edge sharpening are

$$H = \begin{bmatrix} 0 & -1 & 0 \\ -1 & 5 & -1 \\ 0 & -1 & 0 \end{bmatrix} \quad H = \begin{bmatrix} -1 & -1 & -1 \\ -1 & 9 & -1 \\ -1 & -1 & -1 \end{bmatrix} \quad H = \begin{bmatrix} 1 & -2 & 1 \\ -2 & 5 & -2 \\ 1 & -2 & 1 \end{bmatrix}$$

$$(8.26)$$

The summation of the elements in each mask equals 1. The algorithm will be:

(a) Examine all the pixels and compute $g(j,k)$ by convolving $f(j,k)$ with H.

(b) Either use this computed value of $g(j,k)$ as the new gray level of the pixel concerned or use the original value of $f(j,k)$, depending on whether the difference between the computed $g(j,k)$ and the original $f(j,k)$ is greater than or less than the threshold chosen.

Some other edge enhancement masks and operators have been suggested by different authors. These masks can be used to combine with the original image array to yield a two-dimensional discrete differentiation.

Compass gradient masks: These are so named because they indicate the slope directions of maximum responses, as can be seen by the dashed angle in Fig. 8.13.

Laplacian masks: These masks can be used to detect lines, line ends, and points over edges. By convolution of an image with a laplacian mask, we can obtain edge sharpening without regard to edge direction. The mask shown in part (b) of Fig. 8.14 is obtained by adding the mask shown in part (a) to the result obtained when part (a) is rotated by 45°. The mask shown in part (c) is obtained by subtracting the mask in part (a), after it has been rotated by 45°, from 2 times the mask shown in part (a).

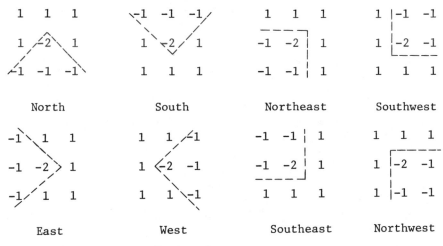

FIG. 8.13 Compass gradient masks.

Gradient masks: The Roberts cross gradient is a 2 × 2 mask. See Eq. (8.24) for the implementation.

Nonlinear edge operators: Among the nonlinear edge operators are the Sobel operator, Kirsch operator, and Wallis operator. Sobel operator uses a 3 × 3 window centered at f(j,k), as shown in Fig. 8.15. The gradient at the point f(j,k) is defined as either

$$S = (s_x^2 + s_y^2)^{1/2} \qquad (8.28)$$

or

$$S = |S_x| + |S_y| \qquad (8.29)$$

where S_x and S_y are, respectively, computed from its neighbor according to

0	-1	0		-1	-1	-1		1	-2	1
-1	4	-1		-1	8	-1		-2	4	-2
0	-1	0		-1	-1	-1		1	-2	1

(a)	(b)	(c)

FIG. 8.14 Laplacian masks.

$$A_0 \quad A_1 \quad A_2$$

$$A_7 \quad f(j, k) \quad A_3$$

$$A_6 \quad A_5 \quad A_4$$

FIG. 8.15 A 3 × 3 window for edge detection.

$$S_x = (A_2 + 2A_3 + A_4) - (A_0 + 2A_7 + A_6) \tag{8.30}$$

and

$$S_y = (A_6 + 2A_5 + A_4) - (A_0 + 2A_1 + A_2) \tag{8.31}$$

Another 3 × 3 nonlinear edge enhancement algorithm was suggested by Kirsch (see Fig. 8.15). The subscripts of its neighbors are so made that they are labeled in ascending order. Modulo 8 arithmetic is used in this computation. Then the enhancement value of the pixel is given as

$$G(j,k) = \max \left(1, \max_{i=0}^{7} [\, |5S_i - 3T_i| \,] \right) \tag{8.32}$$

where S_i and T_i are computed, respectively, from

$$S_i = A_i + A_{i+1} + A_{i+2} \tag{8.33}$$

and

$$T_i = A_{i+3} + A_{i+4} + A_{i+5} + A_{i+6} + A_{i+7} \tag{8.34}$$

Take, for example, an illustrative window as shown in Fig. 8.16. When $i = 2$,

$$S_2 = A_2 + A_3 + A_4 = 3$$

$$T_2 = A_5 + A_6 + A_7 + A_0 + A_1 = 17$$

5	1	1
5	1	1
5	1	1

FIG. 8.16 Illustrative window.

and

$$\left| 5S_2 - 3T_2 \right| = 36$$

Similarly, we can get eight values for $\left| 5S_i - 3T_i \right|$, i = 0, ..., 7, as follows:

i	S_i	T_i	$\left| 5S_i - 3T_i \right|$
0	7	13	4
1	3	17	36
2	3	17	36
3	3	17	36
4	7	13	4
5	11	9	28
6	15	5	60
7	11	9	28

Substitution of these values into Eq. (8.32) gives the gradient at the point G(j,k) = 60. It is not difficult to see from Eq. (8.32) that when the window is passed into a smoothed region (i.e., with the same gray level in the neighbors as the center pixel), $\left| 5S_i - 3T_i \right| = 0$, i = 0, ..., 7. Then the gradient G(j,k) at this center pixel will assume a value of 1. Basically, the Kirsch operator provides the maximal compass gradient magnitude about an image point [ignoring the pixel value of f(j,k)].

A logarithmic laplacian edge detector has been suggested by Wallis. The principle of this detector is based on the homomorphic image processing. The assumption that Wallis made is that if the difference between the absolute value of the logarithm of the pixel luminance and the absolute value of the average of the logarithms of its four nearest neighbors is greater than a threshold value, an edge is assumed to exist.

Using the same window as that used by Kirsch (Fig. 8.15), an expression for G(j,k) can be put in the following form:

$$G(j,k) = \log[f(x,y)] - \frac{1}{4} \log[A_1 A_3 A_5 A_7] \qquad (8.35)$$

or

$$G(j,k) = \frac{1}{4} \log \frac{[f(x,y)]^4}{A_1 A_3 A_5 A_7} \tag{8.36}$$

It can be seen that the logarithm does not have to be computed when compared with the threshold. The computation will be simplified. In addition, this technique is insensitive to multiplicative change in the luminance level, since $f(x,y)$ changes with A_1, A_3, A_5, and A_7 by the same ratio.

8.1.3 Enhancement Through Clustering Specification

Although lots of approaches have been suggested, to date, no general procedure can be followed for image enhancement. Approaches available for image enhancement are very problem oriented. Nevertheless, a more-or-less generalized approach for image enhancement is still being sought. An approach by clustering specification inspired from bionics has been suggested by Bow and Toney (1983). The basics of this approach is quite intuitive. This follows from what we generally expect on a processed image:

1. *Object—distinctive*: It is expected that all the desirous objects should be included in the processed image and, in addition, those separate objects should be as distinct from one another as possible.
2. *Details—discernible*: Fine details of the desirous objects are expected to be discernible as well as possible.

That is, in viewing and in analyzing an image, the first thing to do is usually to separate the objects from the whole image and then focus our attention on the details of each of the objects. Following such bionic requirements, this algorithm consists of first applying the natural clustering method to identify the objects and then allocating appropriate dynamic ranges for each individual object in the order of their importance, so as to be able to fully utilize the gray levels to delineate the details of the objects of interest, leaving the other objects, such as the background, suppressed in the picture. Any clustering approach can be used for clustering purposes. Two of these have been used for illustration: the extreme point clustering approach and the ISODATA algorithm.

Figure 8.17b shows the image after processing of the original image, shown in Fig. 8.17a. Remember that the same image has been studied in Chap. 4. By using this method, a general routine procedure can be followed and the same result obtained with much less human intervention.

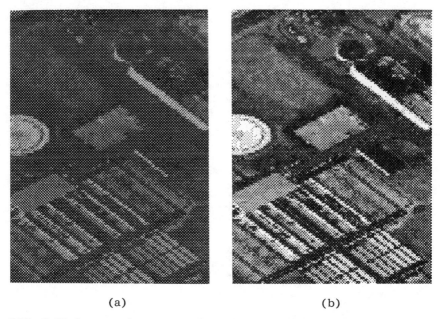

(a)   (b)

FIG. 8.17 Image enhancement through clustering specification: (a) original image; (b) image after processing.

8.2 ENHANCEMENT BY TRANSFORM PROCESSING

As discussed at the beginning of this chapter, image enhancement can also be carried out by the transform method. In this method the image $f(x,y)$ is first transformed into $F(u,v)$, and then processed in the transform domain to meet our requirements. Filtering is one of the processes most frequently used in the transform domain. Since convolution in the spatial domain is converted to simpler multiplication in the transform domain, the processing work required is greatly simplified. On the other hand, extra work will be introduced in the transform and inverse transform of the image function to yield the final spatial image that is expected. A trade-off between these two is therefore needed in making the choice as to the domain in which we are going to work.

The whole procedure in transform processing can be put in block form as shown in Fig. 8.18; where $f(x,y)$ and $\hat{f}(x,y)$ represent, re-

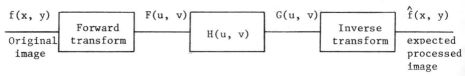

FIG. 8.18 Block diagram of image processing in the transform domain.

spectively, the original image and the expected processed image.
$H(u,v)$ is the process expected to be used in the transform domain,
and $G(u,v)$ is the result after processing in the transform domain,
which can be represented by

$$G(u,v) = F(u,v)H(u,v) \tag{8.37}$$

Filtering is one of the processes used to advantage in the transform
domain. Digital filtering can be implemented ideally in the transform
domain because only pure mathematics is involved rather than physical
components.

Various kinds of filters are available. They can be grouped into
two main categories: low-pass filters and high-pass filters. As we
know, the high-frequency information content of the spectrum (say,
a Fourier spectrum) is contributed primarily by the edges and sharp
transitions, while the low-frequency information content is contributed
by the brightness and the image texture. Depending on what we expect of the processed images, either high-pass or low-pass filters will
be chosen to fit the requirements.

Among the low-pass filters, there are for conventional use the
ideal low-pass filters, Butterworth filters, exponential low-pass
filters, trapezoidal filters, and others. For the ideal low-pass filter
shown in Fig. 8.19a, the transfer function is

$$H(u,v) = \begin{cases} 1 & \text{if } D(u,v) \leq D_0 \\ \\ 0 & \text{if } D(u,v) > D_0 \end{cases} \tag{8.38}$$

where D is the distance from point (u,v) to the origin of the frequency
plane such that $D = (u^2 + v^2)^{1/2}$, and D_0 is the cutoff frequency, a
specified nonnegative quantity, the value of which depends on what
we required in the processed image. D_0 may be obtained by decreasing D until the energy passed exceeds a prescribed percentage of the
total energy.

For the Butterworth low-pass filter shown in Fig. 8.19b, the
transfer function is

$$H(u, v) = \frac{1}{1 + [D(u,v)/D_0]^{2n}} \tag{8.39}$$

where n is known as the order of the filter, and D_0 is the cutoff frequency, which is defined at the point on the abscissa where $H(u,v)$ is
equal to one-half of its maximum value. The image processed with this
Butterworth filter will be expected to have less blurring effect, since
some of the high-frequency-component information will be included in

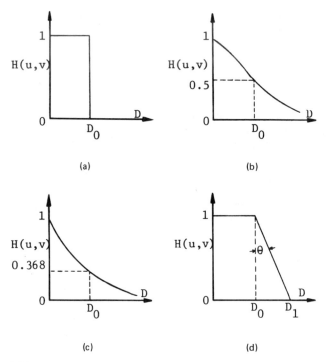

FIG. 8.19 Low-pass filters: (a) ideal; (b) Butterworth; (c) exponential; (d) trapezoidal.

the tail region of this filter, as can be seen in Fig. 8.19b.

The exponential low-pass filter is shown in Fig. 8.19c. Its
transfer function is

$$H(u,v) = e^{-[D(u,v)/D_0]^n} \qquad (8.40)$$

The cutoff frequency D_0 is defined at the point on the abscissa where
$H(u,v)$ drops to a point equal to 0.368 of its maximum value. n in the
transfer function is the variable controlling the rate of decay of the H
function. More blurring will be expected from the exponential low-
pass filter than from the Butterworth filter, since less high-frequency
component information is included in the processed image.

A trapezoidal filter as shown in Fig. 8.19d is halfway between an
ideal low-pass filter and a completely smooth filter. Depending on the
slope of the tail of this trapezoid, the high-frequency-component information content will be different, and therefore the blurring effects
will be different for different cases. From the functional diagram
shown in Fig. 8.19d, $H(u,v)$ assumes a value

$$1 - (D - D_0) \cot \theta = \frac{D_1 - D}{D_1 - D_0}$$

when D falls at a point between D_0 and D_1. Thus we obtain the transfer function of the trapezoidal filter as

$$H(u,v) = \begin{cases} 1 & \text{if } D < D_0 \\ \dfrac{D - D_1}{D_0 - D_1} & \text{if } D_0 \leqslant D \leqslant D_1 \\ 0 & \text{if } D > D_1 \end{cases} \qquad (8.41)$$

Similar to the low-pass filters, we have the ideal high-pass filters, Butterworth high-pass filters, exponential high-pass filters, and trapezpidal high-pass filters. As implied by the name "high-pass," the lower-frequency-component information is attenuated without disturbing the high-frequency information. These kinds of filters are generally used to achieve edge sharpening. The transfer functions for these filters are shown in Fig. 8.20. Contrary to the transfer function shown by Eq. (8.38), the transfer function of an ideal high-pass filter is given by the following relation:

$$H(u,v) = \begin{cases} 0 & \text{if } D < D_0 \\ 1 & \text{if } D > D_0 \end{cases} \qquad (8.42)$$

where $D = D(u,v) = (u^2 + v^2)^{1/2}$. That of the Butterworth high-pass filter is

$$H(u,v) = \frac{1}{1 + [D_0/D(u,v)]^{2n}} \qquad (8.43)$$

Note the difference between this expression and Eq. (8.39). Values of $H(u,v)$ increase with an increase in $D(u,v)$. $H(u,v) = 0$ when $D(u,v)$ is very small; $H(u,v) = 1$ when $D(u,v)$ is much greater than D_0; and $H(u,v) = 0.5$ when $D(u,v) = D_0$ and $n = 1$.

For exponential high-pass filters, the transfer function will be represented by

$$H(u,v) = e^{-[D_0/D(u,v)]^n} \qquad (8.44)$$

$H(u,v)$ is zero when $D(u,v) = 0$, and the cutoff frequency D_0 is then defined at the point on the abscissa when $H(u,v) = e^{-1}$ or 0.368 of the

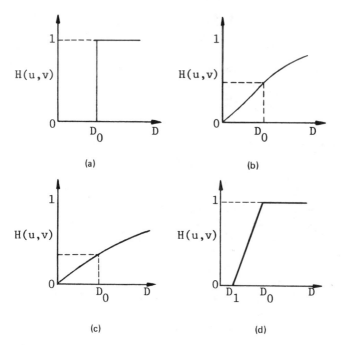

FIG. 8.20 High-pass filters: (a) ideal; (b) Butterworth; (c) exponential; (d) trapezoidal.

maximum value of $H(u,v)$. $H(u,v)$ increases as D increases, and equals 1 when D approaches ∞ as the limit. That is, more high-frequency-component information will be included in the processing, but low-frequency-component information will be suppressed.

Analogous arguments can be applied to the high-pass trapezoidal filter. The transfer function of this filter can be similarly derived as follows:

$$H(u,v) = \begin{cases} 0 & \text{if } D < D_1 \\ \dfrac{D - D_1}{D_0 - D_1} & \text{if } D_0 \geqslant D \geqslant D_1 \\ 1 & \text{if } D > D_0 \end{cases} \qquad (8.45)$$

Comparison of these four transfer functions shows that the high-frequency-component emphasis increases in the order: Butterworth high-pass filter, exponential high-pass filter, and trapezoidal high-pass filter, but the preservation of low-frequency information is in the reverse order for these filters. The proper choice of filter is largely problem dependent.

PROBLEMS

8.1 Use the data of Fig. A.2 of Appendix A. Alter the data by processing each pixel in the image with the following deterministic gray-level transformations, where 0 = dark and 255 = white.

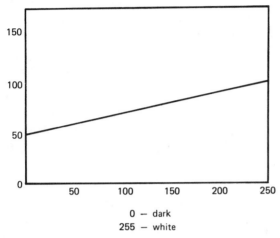

0 — dark
255 — white

8.2 Use the altered data obtained from Problem 8.1 as an example of image data for enhancement processing.

   (a) Obtain the histogram of the example image.

   (b) Obtain a processed image by applying the histogram equalization algorithm to these example image data.

   (c) Evaluate the histogram equalization algorithm by comparing the example image with the image after histogram equalization.

   (d) Suggest a histogram specification transformation and see whether this can further improve the image.

8.3 Use the same example image data. Obtain an edge enhancement image with

$$g(x,y) = \begin{cases} G[f(x,y)] & \text{if } G[f(x,y)] \geqslant T \\ f(x,y) & \text{otherwise} \end{cases}$$

where $G[f(x,y)]$ = the gradient of f at the point (x,y) and T is a nonnegative threshold. Compare the image after processing with the original image and note the improvement in image quality.

8.4 Repeat Problems 8.1 to 8.3 for the chest x-ray image data of Fig. A.7. The image after processing will be much clearer than the corrupted example image, which is usually the image obtained from the x-ray machine. It can easily be seen that the image after processing is of enormous help in medical diagnosis, which is one application of pattern recognition.

appendix A
Digitized Images

The following seven 512 × 512, 256-gray-level images can be used as
large data sets for many of the pattern recognition and data preprocessing concepts developed in the text. These data sets can be used
in their original form or they can be corrupted: for example, by
adding noise to each pixel or by any specified gray-level transformation. By so doing, a variety of input data sets can be generated which
can be used to illustrate the algorithms for pattern recognition as well
as for data preprocessing. The results obtained can be displayed on
a standard line printer or on a dot printer, as shown in the text.
These digitized images were recorded on an IBM multitape, which can
be obtained from the author.

FIG. A.1 A 512 × 512 256 gray-level digitized image.

FIG. A.2 A 512 × 512 256 gray-level digitized image.

FIG. A.3   A 512 × 512 256 gray-level digitized image.

FIG. A.4 A 512 × 512 gray-level digitized image.

FIG. A.5   A 512 × 512 256 gray-level digitized image.

FIG. A.6 A 512 × 512 256 gray-level digitized image.

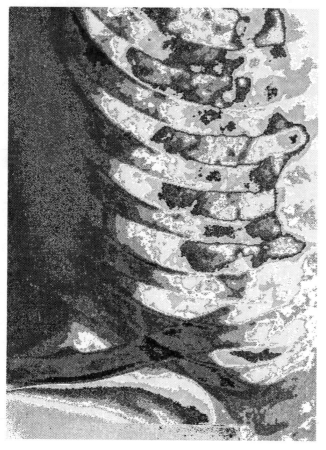

FIG. A.7   A 512 × 512 256 gray-level digitized image.

appendix B
Matrix Manipulation

B.1 DEFINITION

A rectangular array

$$\underline{A} = [a_{ij}]_{m \times n} = \begin{bmatrix} a_{11} & a_{12} & \cdots & a_{1n} \\ a_{21} & a_{22} & \cdots & a_{2n} \\ \cdot & & & \\ \cdot & & & \\ \cdot & & & \\ a_{m1} & a_{m2} & \cdots & a_{mn} \end{bmatrix} \quad (B.1)$$

with a_{ij}, $i = 1, \ldots, m$; $j = 1, \ldots, n$, as the elements is called a matrix of m rows and n columns, or in brief, an *m × n matrix*. The rows

$$\alpha_i = (a_{i1}, \ldots, a_{in}) \quad i = 1, \ldots, m \quad (B.2)$$

are called the *row vectors* of A. Similarly, the columns

$$\beta_j = (a_{1j}, \ldots, a_{mj}) \quad j = 1, \ldots, n \quad (B.3)$$

are called the *column vectors* of A.

A transposed matrix $\underline{A}^T$ is obtained from matrix $\underline{A}$ by interchanging its rows and columns. Thus

$$\underline{A}^T = \begin{bmatrix} a_{11} & a_{21} & \cdots & a_{m1} \\ a_{12} & a_{22} & & a_{m2} \\ \cdot & & & \cdot \\ \cdot & & & \cdot \\ \cdot & & & \cdot \\ a_{1n} & a_{2n} & \cdots & a_{mn} \end{bmatrix} \tag{B.4}$$

If $m = n$, the matrix $\underline{A}$ is said to be a *square matrix*. The main diagonal of this $n \times n$ matrix $\underline{A}$ consists of the entries a_{11}, a_{22}, ..., a_{nn}. If all other entries of $\underline{A}$ are zero, $\underline{A}$ is a *diagonal matrix*. A diagonal matrix $\underline{A}$ is a scalar matrix if $a_{11} = a_{22} = \cdots = a_{nn}$, and is the $n \times n$ identity matrix if $a_{11} = a_{22} = \cdots = a_{nn} = 1$.

A *lower triangular matrix* is a square matrix having all elements above the main diagonal zero. An *upper triangular matrix* can be defined similarly.

B.2 MATRIX MULTIPLICATION

Let $\underline{A}$ be an $m \times p$ matrix and $\underline{B}$ be an $p \times n$ matrix; the product

$$\underline{C} = \underline{A}\underline{B} \tag{B.5}$$

is an $m \times n$ matrix with (i,j)th element represented by

$$c_{ij} = \sum_{k=1}^{p} a_{ik} b_{kj} \tag{B.6}$$

Note that the two matrices should be *conformable*. That is, the two matrices may be multiplied only if the number of columns of the first equals the number of rows of the second.

Premultiplying a matrix by a conformable diagonal matrix has the effect of scaling each row of the matrix by the corresponding element in the diagonal matrix:

$$\begin{bmatrix} a_{11} & & 0 \\ & a_{22} & \\ 0 & & a_{33} \end{bmatrix} \begin{bmatrix} b_{11} & b_{12} \\ b_{21} & b_{22} \\ b_{31} & b_{32} \end{bmatrix} = \begin{bmatrix} a_{11}b_{11} & a_{11}b_{12} \\ a_{22}b_{21} & a_{22}b_{22} \\ a_{33}b_{31} & a_{33}b_{32} \end{bmatrix} \tag{B.7}$$

In a similar manner, postmultiplying a matrix by a conformable diagonal matrix has the effect of scaling each column by the corresponding element in the diagonal matrix. Thus

$$
\begin{bmatrix} a_{11} & a_{12} & a_{13} \\ a_{21} & a_{22} & a_{23} \\ a_{31} & a_{32} & a_{33} \end{bmatrix} \begin{bmatrix} b_{11} & & 0 \\ & b_{22} & \\ 0 & & b_{33} \end{bmatrix} = \begin{bmatrix} a_{11}b_{11} & a_{12}b_{22} & a_{13}b_{33} \\ a_{21}b_{11} & a_{22}b_{22} & a_{23}b_{33} \\ a_{31}b_{11} & a_{32}b_{22} & a_{33}b_{33} \end{bmatrix}
$$

$$(B.8)$$

The product of two like triangular matrices produces a third matrix of like form, such as

$$
\begin{bmatrix} a_{11} & & \\ a_{21} & a_{22} & \\ a_{31} & a_{32} & a_{33} \end{bmatrix} \begin{bmatrix} b_{11} & & \\ b_{21} & b_{22} & \\ b_{31} & b_{32} & b_{33} \end{bmatrix}
$$

$$
= \begin{bmatrix} a_{11}b_{11} & & \\ a_{21}b_{11} + a_{22}b_{21} & a_{22}b_{22} & \\ a_{31}b_{11} + a_{32}b_{21} + a_{33}b_{31} & a_{32}b_{22} + a_{33}b_{32} & a_{33}b_{33} \end{bmatrix} \quad (B.9)
$$

B.3 PARTITIONING OF MATRICES

Partitioning of matrices by rows and/or columns is helpful in the manipulation of matrices that are more complex in nature or large in size. A simple example is given here for illustration. Suppose that we have a complex simultaneous equations as follows:

$$
\begin{bmatrix} 5+i & 3-i \\ 6-2i & 8+4i \end{bmatrix} \begin{bmatrix} x_1 \\ x_2 \end{bmatrix} = \begin{bmatrix} 6 \\ 5-5i \end{bmatrix} \tag{B.10}
$$

The set of equations above may be put in the following form:

$$
[\underline{A}_r + i\underline{A}_i][\underline{x}_r + i\underline{x}_i] = \underline{b}_r + i\underline{b}_i \tag{B.11}
$$

or

$$
\underline{A}_r\underline{x}_r - \underline{A}_i\underline{x}_i = \underline{b}_r \tag{B.12}
$$
$$
\underline{A}_i\underline{x}_r + \underline{A}_r\underline{x}_i = \underline{b}_i
$$

which may be put in a supermatrix form as

$$
\begin{bmatrix} \underline{A}_r & -\underline{A}_i \\ \underline{A}_i & \underline{A}_r \end{bmatrix} \begin{bmatrix} \underline{x}_r \\ \underline{x}_i \end{bmatrix} = \begin{bmatrix} \underline{b}_r \\ \underline{b}_i \end{bmatrix}
$$ (B.13)

where

$$
\underline{A}_r = \begin{bmatrix} 5 & 3 \\ 6 & 8 \end{bmatrix} \qquad \underline{A}_i = \begin{bmatrix} 1 & -1 \\ -2 & 4 \end{bmatrix}
$$

$$
\underline{b}_r = \begin{bmatrix} 6 \\ 5 \end{bmatrix} \qquad \underline{b}_i = \begin{bmatrix} 0 \\ -5 \end{bmatrix}
$$

and $\underline{x}_r$ and $\underline{x}_i$ are column vectors constituting the real and imaginary parts of the solution vector. A set of complex simultaneous equations can then be converted into a set of real simultaneous equations.

Substituting the numerical values for $\underline{A}_r$, $\underline{A}_i$, $\underline{b}_r$, and $\underline{b}_i$ in Eq. (B.13), we obtain

$$
\begin{bmatrix} \begin{vmatrix} 5 & 3 \\ 6 & 8 \end{vmatrix} & - & \begin{vmatrix} 1 & -1 \\ -2 & 4 \end{vmatrix} \\ \begin{vmatrix} 1 & -1 \\ -2 & 4 \end{vmatrix} & & \begin{vmatrix} 5 & 3 \\ 6 & 8 \end{vmatrix} \end{bmatrix} \begin{bmatrix} x_r \\ x_i \end{bmatrix} = \begin{bmatrix} \begin{vmatrix} 6 \\ 5 \end{vmatrix} \\ \begin{vmatrix} 0 \\ -5 \end{vmatrix} \end{bmatrix}
$$ (B.14)

When

$$
\underline{x}_r = \begin{bmatrix} x_1^r \\ x_2^r \end{bmatrix} \qquad \text{and} \qquad x_i = \begin{bmatrix} x_1^i \\ x_2^i \end{bmatrix}
$$

are used, Eq. (B.14) becomes

$$
\begin{bmatrix} 5 & 3 \\ 6 & 8 \end{bmatrix} \begin{bmatrix} x_1^r \\ x_2^r \end{bmatrix} - \begin{bmatrix} 1 & -1 \\ -2 & 4 \end{bmatrix} \begin{bmatrix} x_1^i \\ x_2^i \end{bmatrix} = \begin{bmatrix} 6 \\ 5 \end{bmatrix}
$$ (B.15)

$$
\begin{bmatrix} 1 & -1 \\ -2 & 4 \end{bmatrix} \begin{bmatrix} x_1^r \\ x_2^r \end{bmatrix} + \begin{bmatrix} 5 & 3 \\ 6 & 8 \end{bmatrix} \begin{bmatrix} x_1^i \\ x_2^i \end{bmatrix} = \begin{bmatrix} 0 \\ -5 \end{bmatrix}
$$

or

$$
\begin{bmatrix} 5x_1^r + 3x_2^r \\ 6x_1^r + 8x_2^r \end{bmatrix} - \begin{bmatrix} x_1^i - x_2^i \\ -2x_1^i + 4x_2^i \end{bmatrix} = \begin{bmatrix} 6 \\ 5 \end{bmatrix}
$$

$$
\begin{bmatrix} x_1^r - x_2^r \\ -2x_1^r + 4x_2^r \end{bmatrix} + \begin{bmatrix} 5x_1^i + 3x_2^i \\ 6x_1^i + 8x_2^i \end{bmatrix} = \begin{bmatrix} 0 \\ -5 \end{bmatrix}
$$

(B.16)

Solving the set of simultaneous equations above, we obtain

$$
x_1^r = 1.26 \qquad x_2^r = -0.22
$$

$$
x_1^i = -0.32 \qquad x_2^i = 0.04
$$

or

$$
\underline{x}_1 = x_1^r + ix_1^i = 1.26 - i0.32
$$

$$
\underline{x}_2 = x_2^r + ix_2^i = -0.22 + i0.04
$$

B.4 COMPUTATION OF THE INVERSE MATRIX

In general, a *determinant* can be evaluated as

$$
\det \underline{A} = \sum_{i=1}^{n} (-1)^{i+j} a_{ij} \det \underline{A}(i \,|\, j)
$$

(B.17)

where a_{ij} is the (i,j)th entry of $\underline{A}$ and $\underline{A}(i \,|\, j)$ is an $(n - 1) \times (n - 1)$ matrix obtained by deleting the ith row and jth column of $\underline{A}$. The scalar $(-1)^{i+j} \det \underline{A}(i \,|\, j)$ is called the *cofactor* of the (i,j)th entry of $\underline{A}$ and is denoted simply by c_{ij}. Equation (B.17) then becomes

$$\det \underline{A} = \sum_{i=1}^{n} a_{ij} c_{ij} \tag{B.18}$$

In computing a specific determinant, it is frequently easier step by step to reduce the determinant to a size of lower order. This can be done by adding a multiple of one row of the determinant to another (or a multiple of one column to another).

Take (B.15) as an example. x_1^r can be solved as

$$x_1^r = \frac{\begin{vmatrix} 6 & 3 & -1 & 1 \\ 5 & 8 & 2 & -4 \\ 0 & -1 & 5 & 3 \\ -5 & 4 & 6 & 8 \end{vmatrix}}{|A|} \tag{B.19}$$

where

$$|A| = \begin{vmatrix} 5 & 3 & -1 & 1 \\ 6 & 8 & 2 & -4 \\ 1 & -1 & 5 & 3 \\ -2 & 4 & 6 & 8 \end{vmatrix} \tag{B.20}$$

or

$$|A| = \begin{vmatrix} 1 & -1 & 5 & 3 \\ 6 & 8 & 2 & -4 \\ 5 & 3 & -1 & 1 \\ -2 & 4 & 6 & 8 \end{vmatrix} \tag{B.21}$$

By subtracting suitable multiples of row 1 from rows 2, 3, and 4, respectively, $|A|$ can be reduced to

$$|A| = -\begin{vmatrix} 1 & -1 & 5 & 3 \\ 0 & 14 & -28 & -22 \\ 0 & 8 & -26 & -14 \\ 0 & 2 & 16 & 14 \end{vmatrix} \tag{B.22}$$

or

$$|A| = -8\begin{vmatrix} 7 & -14 & -11 \\ 4 & -13 & -7 \\ 1 & 8 & 7 \end{vmatrix} \tag{B.23}$$

By following the same procedure, the determinant can be further reduced by

$$|A| = -400 \begin{vmatrix} 0 & 7 & 6 \\ 0 & 9 & 7 \\ 1 & 8 & 7 \end{vmatrix} \qquad (B.24)$$

or

$$|A| = -400 \begin{vmatrix} 7 & 6 \\ 9 & 7 \end{vmatrix} = 2000$$

The *adjoint* of $\underline{A}$ is defined as an $n \times n$ matrix, which is the transpose of the matrix of cofactors of $\underline{A}$. Thus

$$(\text{adj } \underline{A})_{ij} = c_{ji} = (-1)^{i+j} \det \underline{A}(j \,|\, i) \qquad (B.25)$$

The *inverse matrix for* $\underline{A}$ is*

$$\underline{A}^{-1} = (\det \underline{A})^{-1} \text{ adj } \underline{A} = \frac{1}{\det \underline{A}} \text{ adj } \underline{A} \qquad (B.26)$$

Example Use the adjoint formula to compute the inverse of the following 3×3 real matrix:

$$\underline{A} = \begin{bmatrix} \cos \theta & 0 & -\sin \theta \\ 0 & 1 & 0 \\ \sin \theta & 0 & \cos \theta \end{bmatrix}$$

Solution: The determinant of the matrix $\underline{A}$ is

$$\det \underline{A} = \cos^2 \theta + \sin^2 \theta = 1$$

The adjoint of $\underline{A}$ is

---

*For the derivation of this relation, see Hoffman and Kunze (1971).

$$\text{adj } \underline{A} = \begin{bmatrix} \begin{vmatrix} 1 & 0 \\ 0 & \cos\theta \end{vmatrix} & -\begin{vmatrix} 0 & -\sin\theta \\ 0 & \cos\theta \end{vmatrix} & \begin{vmatrix} 0 & -\sin\theta \\ 1 & 0 \end{vmatrix} \\ -\begin{vmatrix} 0 & 0 \\ \sin\theta & \cos\theta \end{vmatrix} & \begin{vmatrix} \cos\theta & -\sin\theta \\ \sin\theta & \cos\theta \end{vmatrix} & -\begin{vmatrix} \cos\theta & -\sin\theta \\ 0 & 0 \end{vmatrix} \\ \begin{vmatrix} 0 & 1 \\ \sin\theta & 0 \end{vmatrix} & -\begin{vmatrix} \cos\theta & 0 \\ \sin\theta & 0 \end{vmatrix} & \begin{vmatrix} \cos\theta & 0 \\ 0 & 1 \end{vmatrix} \end{bmatrix}$$

$$= \begin{vmatrix} \cos\theta & 0 & \sin\theta \\ 0 & 1 & 0 \\ -\sin\theta & 0 & \cos\theta \end{vmatrix}$$

Therefore,

$$A^{-1} = \frac{1}{\det \underline{A}} \text{ adj } \underline{A} = \begin{vmatrix} \cos\theta & 0 & \sin\theta \\ 0 & 1 & 0 \\ -\sin\theta & 0 & \cos\theta \end{vmatrix}$$

appendix C
Eigenvectors and Eigenvalues of an Operator

An eigenvalue and corresponding eigenvector of a matrix satisfy the property that the eigenvector multiplied by the matrix yields a vector proportional to itself. In other words, an eigenvalue (or characteristic number) of an operator $\underline{A}$ is a number λ such that for some nonzero vector $\underline{x}$ the following equality holds:

$$\underline{A}\underline{x} = \lambda\underline{x} \tag{C.1}$$

where $\underline{x}$, any nonzero vector that satisfies Eq. (C.1), is called an *eigenvector* of the operator $\underline{A}$, and λ, the proportionality constant, is known as the *eigenvalue*. For Eq. (C.1) to be conformable, $\underline{A}$ must be square. Hence only square matrices have eigenvalues. The spectrum of an operator is the set of all its eigenvalues.

Let an operator $\underline{A}$ be represented by

$$\underline{A} = (a_{ik}) \tag{C.2}$$

Then

$$\underline{A}\underline{x} = \sum_{k=1}^{n} a_{1k}x_k, \ldots, \sum_{k=1}^{n} a_{nk}x_k \tag{C.3}$$

Equation (C.1) represents a system of equations as follows:

$$
\begin{vmatrix} a_{11} & a_{12} & \cdots & a_{1n} \\ a_{21} & a_{22} & \cdots & a_{2n} \\ \cdot \\ \cdot \\ \cdot \\ a_{n1} & a_{n2} & \cdots & a_{nn} \end{vmatrix} \begin{vmatrix} x_1 \\ x_2 \\ \cdot \\ \cdot \\ \cdot \\ x_n \end{vmatrix} = \lambda \begin{vmatrix} x_1 \\ x_2 \\ \cdot \\ \cdot \\ \cdot \\ x_n \end{vmatrix}
\qquad \text{(C.4)}
$$

or

$$
\begin{aligned}
a_{11}x_1 + a_{12}x_2 + \cdots + a_{1n}x_n &= \lambda x_1 \\
a_{21}x_1 + a_{22}x_2 + \cdots + a_{2n}x_n &= \lambda x_2 \\
&\ \ \vdots \\
a_{n1}x_1 + a_{n2}x_2 + \cdots + a_{nn}x_n &= \lambda x_n
\end{aligned}
\qquad \text{(C.5)}
$$

After shifting terms, we obtain

$$
\begin{aligned}
(a_{11} - \lambda) + a_{12}x_2 + \cdots + a_{1n}x_n &= 0 \\
a_{21}x_1 + (a_{22} - \lambda)x_2 + \cdots + a_{2n}x_n &= 0 \\
&\ \ \vdots \\
a_{n1}x_1 + a_{n2}x_2 + \cdots + (a_{nn} - \lambda)x_n &= 0
\end{aligned}
\qquad \text{(C.6)}
$$

This system of equations will have a nonzero solution if and only if the following characteristic equation is satisfied:

$$
\begin{vmatrix}
a_{11} - \lambda & a_{12} & \cdots & a_{1n} \\
a_{21} & a_{22} - \lambda & \cdots & a_{2n} \\
\cdot \\
\cdot \\
\cdot \\
a_{n1} & a_{n2} & \cdots & a_{nn} - \lambda
\end{vmatrix} = 0
\qquad \text{(C.7)}
$$

That is, this system of equations will have a nonzero solution if and only if λ is a root of the characteristic polynomial of the matrix.

Example Let A be a (real) 3×3 matrix

$$\begin{bmatrix} 3 & 1 & -1 \\ 2 & 2 & -1 \\ 2 & 2 & 0 \end{bmatrix}$$

Then the characteristic polynomial for $\underline{A}$ is

$$\begin{vmatrix} \lambda - 3 & -1 & 1 \\ -2 & \lambda - 2 & 1 \\ -2 & -2 & \lambda \end{vmatrix} = \lambda^3 - 5\lambda^2 + 8\lambda - 4 = (\lambda - 1)(\lambda - 2)^2$$

Thus the characteristic values (eigenvalues) for $\underline{A}$ are 1 and 2.

By expanding Eq. (C.7) in terms of λ, the following characteristic equation results:

$$\lambda^n + C_{n-1}\lambda^{n-1} + \cdots + C_1\lambda + C_0 = 0 \tag{C.8}$$

which, in turn, can be factorized into the form

$$(\lambda - \lambda_1)(\lambda - \lambda_2) \cdots (\lambda - \lambda_n) = 0 \tag{C.9}$$

showing that a matrix of order n has n eigenvalues.

Comparing Eq. (C.8) with (C.9) and remembering that the *trace* of a matrix is defined as the sum of the diagonal elements of the matrix, we obtain

$$\begin{aligned} \text{tr}(\underline{A}) &= a_{11} + a_{22} + \cdots + a_{nn} \\ &= -C_{n-1} = \lambda_1 + \lambda_2 + \cdots + \lambda_n \end{aligned} \tag{C.10}$$

that is, the sum of the eigenvalues of a matrix is equal to the trace of the matrix.

Also, from Eqs. (C.7) to (C.9),

$$|\underline{A}| = (-1)^n C_0 = \lambda_1 \lambda_2 \cdots \lambda_n \tag{C.11}$$

that is, the product of the eigenvalues of a matrix equals the determinant of the matrix.

appendix D
Notation

CHAPTER 1

$$\underline{x} = \begin{bmatrix} x_1 \\ x_2 \\ \cdot \\ \cdot \\ \cdot \\ x_n \end{bmatrix}$$

Pattern or pattern vector

n   Number of dimensions of the pattern vector

r   Number of dimensions of the feature vector

Pattern space

$$\underline{X} \begin{bmatrix} \underline{x}_1^T \\ \underline{x}_2^T \\ \cdot \\ \cdot \\ \cdot \\ \underline{x}_m^T \end{bmatrix} = \begin{bmatrix} x_{11} & x_{12} & \cdots & x_{1n} \\ x_{21} & x_{22} & \cdots & x_{2n} \\ \cdot & & & \cdot \\ \cdot & & & \cdot \\ \cdot & & & \cdot \\ x_{m1} & x_{m2} & \cdots & x_{mn} \end{bmatrix}$$

$$z_{-k}^{m} = \begin{bmatrix} z_{k1}^{m} \\ z_{k2}^{m} \\ \cdot \\ \cdot \\ \cdot \\ z_{ki}^{m} \\ \cdot \\ \cdot \\ \cdot \\ z_{kn}^{m} \end{bmatrix}$$

mth prototype in class k

$k = 1, \ldots, M$ — Pattern class index

$m = 1, \ldots, N_k$ — Prototype index

M — Number of pattern classes

ω_k — Pattern class k

N_k — Number of prototypes in the kth class

$d(.)$ — A distance function

a, b, c, d — Primitives (or terminals) used for structural analysis

CHAPTER 2

$d(\underline{x})$ — Decision or discriminant function

$d_k(\underline{x})$ — Values of the discriminant function for patterns $\underline{x}$ in class k

$$\underline{w} = \begin{bmatrix} w_1 \\ w_2 \\ \cdot \\ \cdot \\ \cdot \\ w_n \\ w_{n+1} \end{bmatrix}$$

Augmented weight vector

$$\underline{x} = \begin{bmatrix} x_1 \\ x_2 \\ \cdot \\ \cdot \\ \cdot \\ X_n \\ 1 \end{bmatrix}$$

Augmented pattern vector

$d_i(\underline{x}) = \underline{w}_i^T \underline{x}$

Linear decision function

$D(\underline{x}, \underline{z}_k)$

A metric, euclidean distance of an unknown pattern $\underline{x}$ from $\underline{z}_k$

$\underline{z}_k$

Prototype average (or class center) for class ω_k

N_k

The number of prototypes used to represent the category k

$d_k^m(\underline{x})$

Value of discriminant function for the mth prototype in class k

$\forall$

For all

ε

Belongs to

$\notin$

Does not belong to

max

Maximum

Σ

Summation

$\exists$

There exists

$\nexists$

There does not exist

$\underline{w}_i^T = (w_{i1}, w_{i2}, \ldots, w_{in})$,

$i = 1, \ldots, R$

n-dimensional weight vectors for the various discriminant functions in layered machine

$$\underline{A} = \begin{bmatrix} w_{11} & w_{12} & \cdots & w_{1n} \\ w_{21} & w_{22} & \cdots & w_{2n} \\ \cdot & & & \cdot \\ \cdot & & & \cdot \\ \cdot & & & \cdot \\ w_{n1} & & \cdots & w_{nn} \end{bmatrix}$$

Weight matrix used in quadratic discriminant function

$$\underline{B} = \begin{vmatrix} w_1 \\ w_2 \\ . \\ . \\ . \\ w_n \end{vmatrix}$$

Weight vector used in quadratic discriminant function

$C = w_{n+1}$

Weight constant used in quadratic discriminant function

$\phi(\underline{x})$

ϕ function (or generalized decision function) used for pattern classification

$f_i(\underline{x}),\quad i = 1,\ \ldots$

Linearly independent, real and single-valued functions

$D(N,n)$

Linear dichotomies on N patterns in n-dimensional space

$\psi(\underline{x},\underline{z}_k^m)$

A potential function of $\underline{x}$ and $\underline{z}_k^m$ defined over the pattern space

CHAPTER 3

T

Margin (or threshold)

$\underline{w}(k),\ \underline{w}(k+1)$

Weight vectors at the kth and (k + 1)th correction steps

$\underline{Z} = [\underline{z}_1^1, \underline{z}_2^1,\ \ldots,\ \underline{z}_1^{N_1}, \underline{z}_2^{N_2}]$

Training sequence

$\triangledown$

Gradient operator

$J(\underline{w})$

A criterion function used in weight-adjustment procedure

ρ_k

A positive scale factor, used to set the training step size

$J_p(\underline{w})$

Perceptron criterion function

P

Set of samples misclassified by $\underline{w}(k)$

$J_r(\underline{w})$

Relaxation criterion function

sgn $d_i(\underline{z})$	Equal to +1 or −1 according to $d_i(\underline{z}) \geqslant$ or < 0
N	Total number of prototypes for all classes
N_i	Total number of prototypes for class ω_i
M	Total number of classes
$\underline{e}$	Error vector
$J_s(\underline{w})$	Sum-of-squared-error criterion function
$z^{\#}$	Pseudoinverse or generalized inverse of $\underline{z}$
$\delta b_i(k)$	Adjustment for $b(k)$

CHAPTER 4

$p(\omega_i)$	A priori probability of class ω_i
$p(\underline{x})$	Probability density function of $\underline{x}$, or probability that $\underline{x}$ occurs without regard to category to which it belongs
$p(\underline{x}\|\omega_i)$	Likelihood function of class ω_i
$p(\omega_i\|\underline{x})$	Probability that $\underline{x}$ comes from ω_i
L_{ij}	Loss or penalty due to deciding $\underline{x} \in \omega_j$ when in fact $\underline{x} \in \omega_i$
$r_k(\underline{x})$	Conditional average loss (or conditional average risk of misclassifying $\underline{x}$ as in ω_k
$\delta(k-i)$	Kronecker delta function
$\aleph$	Normal density function
$\underline{m}_k$	Mean vector for class k
$\underline{C}_k$	Covariance matrix for class k

$\phi_k(\underline{x})$ Arbitrary functions used for
                                                          determining the statistical
                                                          discriminant function

c_{ik} Coefficients for use with $\phi_k(\underline{x})$
                                                          in determining the statistical
                                                          discriminant function

CHAPTER 5

$\cup$ Union

$\cap$ Intersection

$\zeta(\underline{x}_i, \underline{x}_j)$ Dissimilarity measure between
                                                          two patterns

α_k Weighting coefficient

σ_{km}^2 Variance of the mth cluster
                                                          in the kth direction

$r(\underline{x}_i, \underline{x}_j)$ Mahalanobis distance from
                                                          $\underline{x}_i$ to $\underline{x}_j$

C^{-1} Inverse of the covariance
                                                          matrix

$d_t(x_i, x_j)$ Tanimoto coefficient

$D_{\alpha\beta}$ Interset distance between two
                                                          separate sets $[\underline{x}_\alpha^i]$ and $[\underline{x}_\beta^j]$,
                                                          $i = 1, 2, \ldots, N_\alpha$; $j = 1, 2,$
                                                          $\ldots, N_\beta$

$D_{\alpha\alpha}$ Intraset distance in the set
                                                          $[\underline{x}_\alpha^i]$, $i = 1, 2, \ldots, N_\alpha$

τ Membership boundary for a
                                                          specified cluster

θ Fractional constant for setting
                                                          the indeterminate region

$\underline{z}_i$ ith cluster

$\underline{z}_i(t + 1)$ ith cluster computed with
                                                          $(t + 1)$ pattern samples as-
                                                          signed to it

$\underline{C}(t + 1)$ Variance computed with $(t + 1)$
                                                          pattern samples

Maximin	Maximum of the minimum distances
$\Omega_k(i)$	A set of k-nearest neighbors of the sample points i, i = 1, 2, ..., N, basing on euclidean distance measure
$P_k(i)$	Potential of the pattern sample point i
$d(i,j)$	Euclidean distance between the sample points i and j
$\xi_k(i)$	A set of sample points k-adjacent to the sample point i
$SIM(m,n)$	Similarity measure used for merging two most similar subclusters
$SIM_1(m,n)$	Represents the difference in density between the cluster and the boundary
$SIM_2(m,n)$	Represents the relative size of the boundary to that of the cluster
$Y_k^{m,n}$	Denotes the set of points which are in the cluster m while, at the same time, their respective k adjacent points are in cluster n
W_m	A set of points contained in subcluster m
$BP_k^{m,n}$	Average of $P_k(i)$ over all i in $Y_k^{m,n}$
$P_k^{sc}(m)$	Average of $P_k(i)$ over all points in cluster m
$N(.)$	The number of elements of the set inside the bracket
φ	Denotes a set of unordered pairs of subclusters
$S_j(k)$	Cluster domain j at the kth iteration

N_j	The number of samples in $S_j(k)$
M	Number of clusters desired
η	Minimum number of samples desired in a cluster
σ_s	Maximum standard deviation allowed
δ	Minimum distance required between clusters
L	Maximum number of pairs of cluster centers which can be lumped
I	Number of iterations allowed
$\underline{z}_{i\ell}$, $\underline{z}_{j\ell}$	Cluster centers to be lumped
$N_{i\ell}$, $N_{j\ell}$	Number of samples in clusters $\underline{z}_{i\ell}$ and $\underline{z}_{j\ell}$
R_{ij}	Clustering parameter suggested by Davies and Bouldin to obtain natural partitionings of the data sets
R_i	Maximum of R_{ij}
$\overline{R}$	The average of the similarity measures of each cluster with its most similar cluster
$\lambda_c(N_c)$	A performance index used by DYNOC to determine the optimal clusters
E_i	"Sampling" or multiple centers or cores
S_i	Cluster domains
$D(E_i, S_i)$	Degree of similarity of E_i to S_i
Q	A criterion function used by the dynamic clusters method
s_{ij}	Element (i,j) of the similarity matrix S used in the graph theoretic method

θ	Threshold distance denoting the similarity between two pattern points
S', S'', S'''	Reduced similarity matrices
$\mathbb{G} = [G_1, G_2, \ldots, G_M]$	Graphs
$\mathbb{R} = [R_1, R_2, \ldots, R_M]$	Regions of influence
GG	Gabriel graph
RNG	Relative neighborhood graph
(p_i, p_j)	Line segment
β	A factor of relative edge consistency
Λ	Diagonal matrix

CHAPTER 6

r_p^2	Mahalanobis distance produced by the p features
d_i^2	Contribution of the ith feature to the Mahalanobis distance
ξ	Positive arbitrary constant (less than 1)
w_{jj}	Feature weighting coefficients
$\overline{D}^2$	Intraset distance for pattern points after transformation
σ_j^2	Sample variance of the components along the x_j coordinate direction
ρ_1	Lagrange multiplier
$\prod_{j=1}^{n}$	Continual product
x_{ijk}	The observation (or the intensity of picture element) in the kth channel for picture element j in scan line i

$\hat{\underline{m}}_{\ell}$	The sample mean vector for the ℓth category
$\hat{\underline{C}}_{\ell}$	The covariance matrix for the ℓth category
$\underline{x}$	The observation vector of dimension n
$\underline{y}$	The transformed vector of dimension p
$\underline{C}$	p × n transformation matrix
$\underline{A}$	The covariance of the means of the categories
$\underline{X}$	n × M matrix of all the category means composed of all mean vectors $\hat{\underline{m}}_{k}$
M	Number of categories
$\underline{n}$	M × 1 vector of the number of observations in the categories
$\underline{N}$	M × M matrix of the number of observations in the categories
n_{ℓ}	The number of observations for category ℓ
$\underline{W}$	Combined covariance matrix for all the categories
I	p × p identity matrix
Λ	Diagonal matrix
BEG	Basic event generation
$\underline{E}_{ik}$	Basic events for class ω_i by means of BEG
F_{ik}	A subset of features for each basic event $\underline{E}_{ik}$
M(. , .)	Merging function
$\underline{R}^{n}$	Pattern space
$\underline{x}_1, \underline{x}_2, \ldots, \underline{x}_{N_i}$	Training samples from class ω_i

$\underline{y}_1, \underline{y}_2, \cdots, \underline{y}_{N_j}$	Training samples from classes other than ω_i
$\text{dist}(\underline{y}_\ell \vert \underline{E}_{ik})$	The distance between $\underline{y}_\ell$ and $\underline{E}_{ik}$
T_A	Threshold, a certain positive number
F_{ik}^ℓ	The sets of effective features for ℓ training samples not in class ω_i

CHAPTER 7

$f(x)$	One-dimensional function
$F(u)$	Fourier transform of $f(x)$
$f(x,y)$	Two-dimensional image model
$F(u,v)$	Fourier transform of $f(x,y)$
$\phi(u,v)$	Phase spectrum of $f(x,y)$
$E(u,v)$	Energy spectrum of $f(x,y)$
$g(x,y;\ u,v)$	Forward transform kernel
$h(x,y;\ u,v)$	Inverse transform kernel
δ	Dirac delta function
$g_c(x,u)$ or $g_{col}(x,u)$	Forward column transform kernel
$g_r(y,v)$ or $g_{row}(y,v)$	Forward row transform kernel
$F_x(x,v) = \displaystyle\int_{-\infty}^{\infty} f(x,y)\, e^{-j2\pi vy}\, dy$	Columnwise transform of $f(x,y)$
$F_y(u,y) = \displaystyle\int_{-\infty}^{\infty} f(x,y)\, e^{-j2\pi ux}\, dx$	Rowwise transform of $f(x,y)$
$S(u)$	A sampling impulse train
$W_N = e^{-j2\pi/N}$	A notation used for simplifying the expressions of the Fourier transform pair used as the kernel for the sequence of N terms

$$W_{N/2} = e^{-j2\pi/(N/2)}$$

Used as the kernel for the sequence of $N/2$ terms

$$G(u) = \sum_{f=0}^{(N/2)-1} f(2r)(W_{N/2})^{ru}$$

$(N/2)$-point discrete Fourier transform

$$H(u) = \sum_{r=0}^{(N/2)-1} f(2r + 1)(W_{N/2})^{ru}$$

$(N/2)$-point discrete Fourier transform

$$G_1(u) = \sum_{\ell=0}^{(N/4)-1} g(2\ell)(W_{N/4})^{\ell u}$$

That part of $G(u)$ for even values of r

$$G_2(u) = \sum_{\ell=0}^{(N/4)-1} g(2\ell + 1)(W_{N/4})^{\ell u}$$

That part of $G(u)$ for odd values of r

$$H_1(u) = \sum_{\ell=0}^{(N/4)-1} h(2\ell)(W_{N/4})^{\ell u}$$

That part of $H(u)$ for even values of r

$$H_2(u) = \sum_{\ell=0}^{(N/4)-1} h(2\ell + 1)(W_{N/4})^{\ell u}$$

That part of $H(u)$ for odd values of r

$$g(x,u) = \frac{1}{N} \prod_{i=0}^{n-1} (-1)^{b_i(x)b_{n-1-i}(u)}$$

Forward transform kernel for Walsh transform

$$b_k(z)$$

kth bit in the binary representation of z with the zeroth bit as the least significant one

$$g(x,u) = \frac{1}{N} (-1)^{\sum_{i=0}^{n-1} b_i(x)b_i(u)}$$

Forward transform kernel for Hadamard transform

$$H_N$$

Hadamard matrix of order N

$$H(u,v)$$

$N \times N$ symmetric Hadamard transformation matrix

$$F_H(u,v)$$

Hadamard transform of $f(x,y)$

$$\underline{x}$$

Original image vector

$$\underline{y}$$

Transformed vector

$$(\underline{x} - \underline{m}_x)$$

Centralized image vector

$$\underline{B}$$

$N^2 \times N^2$ transformation matrix

λ_i 　　　　　　　　　　　　Variance of the ith element of
　　　　　　　　　　　　　　　$\underline{y}$ along eigenvector $\underline{e}_j$

CHAPTER 8

$p_r(r)$ 　　　　　　　　　　　　Probability density function
　　　　　　　　　　　　　　　(or the relative occurrence
　　　　　　　　　　　　　　　frequency)

$$\int_{r_1}^{r_i} p_r(r)\ dr$$ 　　　　Cumulative density function

r 　　　　　　　　　　　　　　Input image intensity or gray
　　　　　　　　　　　　　　　levels normalized within the
　　　　　　　　　　　　　　　range $(0,1)$

n_i 　　　　　　　　　　　　Number of pixels at gray level
　　　　　　　　　　　　　　　r_i

s 　　　　　　　　　　　　　　Output image intensity

$T(r)$ 　　　　　　　　　　　　Intensity transformation func-
　　　　　　　　　　　　　　　tion

$H_r(j)$ 　　　　　　　　　　　Histogram of the input image

$H_s(k)$ 　　　　　　　　　　　Histogram of the image after
　　　　　　　　　　　　　　　gray-level transformation

$f'(x,y)$ 　　　　　　　　　　Hypothetically noise-free input
　　　　　　　　　　　　　　　image

$n(x,y)$ 　　　　　　　　　　The noise

$f(x,y)$ 　　　　　　　　　　The noisy image

$\hat{f}(X,y)$ 　　　　　　　　　Expected processed image

$H(u,v)$ 　　　　　　　　　Process transfer function

$G(u,v)$ 　　　　　　　　　Processed image in transform
　　　　　　　　　　　　　　　domain

ϕ 　　　　　　　　　　　　Cost function

$g(x,y)$ 　　　　　　　　　　The smoothed image

$(G - f)^2$ 　　　　　　　　　Squared difference between
　　　　　　　　　　　　　　　$g(x,y)$ and $f(x,y)$

$$\left(\frac{\partial g}{\partial x}\right)^2 + \left(\frac{\partial g}{\partial y}\right)^2 \qquad\qquad \text{Magnitude of the gradient}$$

$W_{x,y}$ Weights over x, y range

τ Threshold

Ω Rectangular n × m neighborhood defined

$W_{n,m}$ n × m weight array

N   Total number of pixels defined by Ω

$\underline{G}[f(x,y)]$ Vector gradient

H   High-pass mask

Bibliography

BOOKS

Aggarwal, J. K., Duda, R. O., and Rosenfeld, A., eds. (1977). *Computer Methods in Image Analysis*, IEEE Press, New York.

Agrawala, A. K., ed. (1977). *Machine Recognition of Patterns*, IEEE Press, New York.

Ahmed, N., and Rao, K. R. (1975). *Orthogonal Transforms for Digital Signal Processing*, Springer-Verlag, Berlin.

Aho, A. V., and Ullman, J. D. (1972). *The Theory of Parsing, Translation and Compiling*, vol. 1, Prentice-Hall, Englewood Cliffs, N.J.

Anderberg, M. R. (1973). *Cluster Analysis for Applications*, Academic Press, New York.

Anderson, T. W. (1958). *Introduction to Multivariate Statistics*, Wiley, New York.

Andrews, H. C. (1970). *Computer Techniques in Image Processing*, Academic Press, New York.

Andrews, H. C. (1972). *Introduction to Mathematical Techniques in Pattern Recognition*, Wiley, New York.

Andrews, H. C., and Hunt, B. R. (1977). *Digital Image Restoration*, Prentice-Hall, Englewood Cliffs, N.J.

Arbib, M. A. (1972). *The Metaphorical Brain: An Introduction to Cybernetics as Artificial Intelligence and Brain Theory*, Wiley, New York.

Batchelor, B. G. (1974). *Practical Approach to Pattern Classification*, Plenum Press, New York.

Batchelor, B. G., ed. (1978). *Pattern Recognition — Ideas in Practice*, Plenum Press, New York.

Bongard, M. (1970). *Pattern Recognition* (J. K. Hawkins, ed.;
    T. Cheron, trans.), Spartan Books, Washington, D.C.
Brigham, E. (1974). *The Fast Fourier Transform*, Prentice-Hall,
    Englewood Cliffs, N.J.
Cacoullos, T., ed. (1973). *Discriminant Analysis and Applications*,
    Academic Press, New York.
Carlson, P. F. (1977). *Introduction to Applied Optics for Engineers*,
    Academic Press, New York.
Chasen, S. H. (1978). *Geometric Principles and Procedures for Com-
    puter Graphic Applications*, Prentice-Hall, Englewood Cliffs, N.J.
Chen, C. H. (1973). *Statistical Pattern Recognition*, Spartan Books,
    Washington, D.C.
Cheney, E. W. (1966). *Introduction to Approximation Theory*,
    McGraw-Hill, New York.
Cornsweet, T. N. (1970). *Visual Perceptron*, Academic Press, New
    York.
Dainty, J. C., and Shaw, R. (1974). *Image Science*, Academic Press,
    New York.
Dodwell, P. C. (1970). *Visual Pattern Recognition*, Holt, Rinehart
    and Winston, New York.
Dodwell, P. C. (1971). *Perceptual Processing*, Appleton-Century-
    Crofts (Meredith), New York.
Duda, R. O., and Hart, P. E. (1973). *Pattern Classification and
    Scene Analysis*, Wiley, New York.
Everitt, B. (1974). *Cluster Analysis*, Wiley (Halsted Press), New
    York.
Fu, K. S. (1968). *Sequential Methods in Pattern Recognition and
    Machine Learning*, Academic Press, New York.
Fu, K. S. (1971). *Pattern Recognition and Machine Learning*, Plenum
    Press, New York.
Fu, K. S. (1974). *Syntactic Methods in Pattern Recognition*, Academic
    Press, New York.
Fu, K. S., ed. (1976). *Digital Pattern Recognition*, Springer-Verlag,
    Berlin.
Fu, K. S., ed. (1977). *Syntactic Pattern Recognition Applications*,
    Springer-Verlag, Berlin.
Fu, K. S. (1982). *Syntactic Pattern Recognition and Applications*,
    Prentice-Hall, Englewood Cliffs, N.J.
Fukunaga, K. (1972). *Introduction to Statistical Pattern Recognition*,
    Academic Press, New York.
Gilio, W. K. (1978). *Interactive Computer Graphics*, Prentice-Hall,
    Englewood Cliffs, N.J.
Gonzalez, R. C., and Thomason, M. G. (1978). *Syntactic Pattern Re-
    cognition: An Introduction*, Addison-Wesley, Reading, Mass.
Gonzalez, R. C., and Wintz, P. A. (1977). *Digital Image Processing*,
    Addison-Wesley, Reading, Mass.

Goodman, J. W. (1968). *Introduction to Fourier Optics*, McGraw-Hill, New York.

Grasseli, A., ed. (1969). *Automatic Interpretation and Classification of Images*, Academic Press, New York.

Helstrom, C. W. (1968). *Statistical Theory of Signal Detection*, Pergamon Press, Elmsford, N.Y.

Hoffman, K., and Kunze, R. (1971). *Linear Algebra*, Prentice-Hall, Englewood Cliffs, N.J.

Hopcroft, J. E., and Ullman, J. D. (1969). *Formal Languages and Their Relation to Automata*, Addison-Wesley, Reading, Mass.

Huang, T. S., ed. (1975). *Picture Processing and Digital Filtering*, Springer-Verlag, Berlin.

Huang, T. S. (1981). *Two-Dimensional Digital Signal Processing II — Transforms and Median Filters*, Springer-Verlag, Berlin.

Huang, T. S., and Tretiak, O. J., ed. (1972). *Proceedings of the 1969 Symposium on Picture Bandwidth Compressing*, Gordon and Breach, New York.

Institute of Physics, Conference Series 13 (1972). *Machine Perception of Patterns and Pictures*, IP, London.

Jackson, P. C. (1974). *Introduction to Artificial Intelligence*, Mason-Charter, New York.

Julesz, B. (1971). *Foundations of Cyclopian Perception*, University of Chicago Press, Chicago.

Kanal, L., ed. (1968). *Pattern Recognition*, Thompson Books, Washington, D.C.

Kanal, L. N., and Rosenfeld, A., eds. (1981). *Progress in Pattern Recognition*, North-Holland, Amsterdam.

Kaneff, S. (1970). *Picture Language Machines*, Academic Press, New York.

Klinger, A., Fu, K. S., and Kunii, T. L., eds. (1977). *Data Structures, Computer Graphics, and Pattern Recognition*, Academic Press, New York.

Kmoth, D. E. (1968). *The Art of Computer Programming*, vol. 1, Addison-Wesley, Reading, Mass.

Kovalevsky, V. A. (1980). *Image Pattern Recognition*, Springer-Verlag, Berlin.

Kullback, S. (1958). *Information Theory and Statistics*, Wiley, New York.

Lipkin, B. S., and Rosenfeld, A., eds. (1970). *Picture Processing and Psychopictures*, Academic Press, New York.

Loeve, M. (1948). *Fonctions aléatoires de second ordre*, Hermann, Paris, pp. 299-352.

Meisel, W. S. (1972). *Computer-Oriented Approaches to Pattern Recognition*, Academic Press, New York.

Mendel, J. M., and Fu, K. S. (1970). *Adaptive Learning and Pattern Recognition Systems*, Academic Press, New York.

Morrison, D. F. (1976). *Multivariate Statistical Methods*, McGraw-Hill, New York.

Nake, F., and Rosenfeld, A., eds. (1972). *Graphic Language*, North-Holland, Amsterdam.

Nevatia, R. (1977). *Structured Descriptions of Complex Curved Objects from Recognition and Visual Memory*, Springer-Verlag, Berlin.

Newman, W. M., and Sproull, R. F. (1973). *Principles of Interactive Graphics*, McGraw-Hill, New York.

Nilsson, N. J. (1965). *Learning Machines*, McGraw-Hill, New York.

Oppenheim, A. V., and Schaefer, R. W. (1975). *Digital Signal Processing*, Prentice-Hall, Englewood Cliffs, N.J.

Papoulis, A. (1968). *Systems and Transforms with Application in Optics*, McGraw-Hill, New York.

Patrick, E. A. (1972). *Fundamentals of Pattern Recognition*, Prentice-Hall, Englewood Cliffs, N.J.

Pavlidis, T. (1977). *Structural Pattern Recognition*, Springer-Verlag, New York.

Pearson, D. E. (1975). *Transmission and Display of Pictorial Information*, Wiley, New York.

Pratt, W. K. (1978). *Digital Image Processing*, Wiley, New York.

Preston, K., Jr. (1972). *Coherent Optical Computers*, McGraw-Hill, New York.

Reza, F. M. (1961). *An Introduction to Information Theory*, McGraw-Hill, New York.

Rogers, D. F., and Adams, J. A. (1976). *Mathematical Elements for Computer Graphics*, McGraw-Hill, New York.

Rogers, D. F., and Adams, B. (1977). *Principles of Interactive Computer Graphics*, McGraw-Hill, New York.

Rosenblatt, F. (1961). *Principles of Neurodynamics: Perceptions and the Theory of Brain Mechanisms*, Spartan Books, Washington, D.C.

Rosenfeld, A. (1969). *Picture Processing by Computer*, Academic Press, New York.

Rosenfeld, A., and Kak, A. C. (1976). *Digital Picture Processing*, Academic Press, New York.

Sebestyen, G. S. (1962). *Decision Making Processes in Pattern Recognition*, ACM Monograph Series, Macmillan, New York.

Sklansky, J. (1973). *Pattern Recognition*, Dowden, Hutchinson & Ross, Stroudsburg, Pa.

Takanori, O. (1976). *Three-Dimensional Imaging Techniques*, Academic Press, New York.

Tatsuoka, M. (1971). *Multivariate Analysis*, Wiley, New York.

Tippet, J. T., ed. (1965). *Optical and Electro-Optical Information Processing*, MIT Press, Cambridge, Mass.

Tou, J. T., and Gonzalez, R. C. (1974). *Pattern Recognition Principles*, Addison-Wesley, Reading, Mass.

Tsypkin, Y. Z. (1971). *Adaptation and Learning in Automatic Systems*, Academic Press, New York.

Tukey, J. W. (1971). *Exploratory Data Analysis*, Addison-Wesley, Reading, Mass.

Uhr, L., ed. (1966). *Pattern Recognition*, Wiley, New York.

Ullman, J. R. (1973). *Pattern Recognition Techniques*, Crane, Russak, New York.

Watanabe, S. (1969). *Knowing and Guessing*, Wiley, New York.

Watanabe, S., ed. (1969). *Methodologies of Pattern Recognition*, Academic Press, New York.

Watanabe, S. (1972). *Frontiers of Pattern Recognition*, Academic Press, New York.

Weizenbaum, J. (1976). *Computer Power and Human Reason*, Freeman, San Francisco.

Winton, P. H. (1977). *Artificial Intelligence*, Addison-Wesley, Reading, Mass.

Wyszecki, G. W., and Stiles, W. S. (1967). *Color Science*, Wiley, New York.

Young, T. Y., and Calvert, T. W. (1974). *Classification, Estimation and Pattern Recognition*, American Elsevier, New York.

Yu, F. T. S. (1982). *Optical Information Processing*, Wiley, New York.

Zusne, L. (1970). *Visual Perception of Form*, Academic Press, New York.

SPECIAL JOURNAL ISSUES

Special issue on redundancy reduction (1967). *Proc. IEEE*, vol. 55, no. 3.

Special issue on digital communications (1971). *IEEE Trans. Commun. Technol.*, vol. COM-19, no. 6, Part 1.

Special issue on digital pattern recognition (1972). *Proc. IEEE*, vol. 60, no. 12.

Special issue on digital picture processing (1972). *Proc. IEEE*, vol. 60, no. 7.

Special issue on two-dimensional digital processing (1972). *IEEE Trans. Comput.*, vol. C-21, no. 7.

Special issue on digital signal processing (1975). *Proc. IEEE*, vol. 63, no. 4.

Computer Graphics and Image Processing (1972-1978). Vols. 1-5.

Special issue on digital filtering and image processing (1975). *IEEE Trans. Circuits Syst.*, vol. CAS-2.

Advances in Digital Image Processing: Theory, Application, Implementation. Plenum, New York, 1979.

Proceedings 1st International Conference on Pattern Recognition (1973).

Proceedings 2nd International Conference on Pattern Recognition (1974).

Proceedings 3rd International Conference on Pattern Recognition (1976).
Proceedings 4th International Conference on Pattern Recognition (1978).
Proceedings 5th International Conference on Pattern Recognition (1980).
Proceedings 6th International Conference on Pattern Recognition (1982).
Modern Utilization of Infra-red Technology VI (1980). Proceedings,
     Society of Photo-optical Instrumentation Engineers, vol. 253.
     (Conference held at San Diego, Calif., July 31-Aug. 1, 1980.)
Proceedings, Society of Photo-optical Instrumentation Engineers, vol.
     241.
IEEE Computer Society Conference on Pattern Recognition and Image
     Processing, 1978 (Conference held at Chicago, Ill., May 31-
     June 2, 1978).
IEEE Computer Society Conference on Pattern Recognition and Image
     Processing, IEEE, New York, 1981. (Conference held at Dallas,
     Tex., Aug. 3-5, 1981.)
1980 Conference on Pattern Recognition, Pattern Recognition, vol.
     14, nos. 1-6.
Proc. PRIP '82. (IEEE Computer Society Conference on Pattern Re-
     cognition and Image Processing, Las Vegas, Nev., June 14-17,
     1982; IEEE, New York, 1982.)
Pattern Recognition Theory and Applications, Proceedings of the
     NATO Advanced Study Institute (1982), Reidel, Dordrecht, The
     Netherlands, 1982. (Conference held at Oxford, England, Mar.
     29-Apr. 10, 1981.)

PAPERS

Abramson, N., and Braverman, D. (1962). Learning to recognize
     patterns in a random environment, IRE Trans. Inf. Theory,
     vol. IT-8, no. 5, pp. S58-S63.
Aggarwal, R. C., and Burrus, C. S. (1975). Number theoretic
     transforms to implement fast digital convolution, Proc. IEEE,
     vol. 63, pp. 550-560.
Agmon, S. (1954). The relaxation method for linear inequalities, Can.
     J. Math., vol. 6, no. 3, pp. 382-392.
Aizerman, M. A., Braverman, E. M., and Rozonoer, L. I. (1964).
     The method of potential functions in the problem of determining
     the characteristics of a function generator from randomly ob-
     served points, Autom. Remote Control, vol. 25, no. 12,
     pp. 1546-1556.
Aizerman, M. A., Braverman, E. M., and Rozonoer, L. I. (1964).
     Theoretical foundations of the potential function method in pat-
     tern recognition, Autom. Remote Control, vol. 25, no. 6,
     pp. 821-837.
Aizerman, M. A., Braverman, E. M., and Rozonoer, L. I. (1965).
     The Robbins-Monro process and the method of potential functions,
     Autom. Remote Control, vol. 26, no. 11, pp. 1882-1885.

Allen, G. R., Bonrud, L. O., Cosgrove, J. J., and Stone, R. M. (1973). Machine processing of remotely sensed data, *Conf. Proc.*, Purdue University, Oct. 1973, IEEE Cat. 73CHO 834-2GE, pp. 1A25-1A42.

Anderson, T. W., and Bahadur, R. R. (1962). Classification into two multivariate normal distributions with different covariance matrices, *Ann. Math. Stat.*, vol. 33, pp. 420-431.

Andrews, H. C. (1971). Multi-dimensional rotations in feature selection, *IEEE Trans. Comput.*, vol. C-20, no. 9, p. 1045.

Andrews, D. F. (1972). Plots of high-dimensional data, *Biometrics*, vol. 28, pp. 125-136.

Andrews, D. F. (1973). Graphical techniques for high-dimensional data, in *Discriminant Analysis and Applications* (T. Cacoullos, ed.), Academic Press, New York, pp. 37-59.

Andrews, H. C. (1972). Some unitary transformations in pattern recognition and image processing, in *Information Processing*, vol. 71 (C. V. Freiman, ed.), North-Holland, Amsterdam.

Andrews, H. C. (1974). Digital image restoration: a survey, *IEEE Comput.*, vol. 7, no. 5, pp. 36-45.

Andrews, H. C., and Patterson, C. L. (1976). Digital interpolation of discrete images, *IEEE Trans. Comput.*, vol. C-25, no. 2, pp. 196-202.

Arcelli, C., and Levialdi, S. (1971). Concavity extraction by parallel processing, *IEEE Trans. Syst. Man Cybern.*, vol. 1, pp. 349-396.

Arcelli, C., and Levialdi, S. (1972). Parallel shrinking in three dimensions, *Comput. Graphics Image Process.*, vol. 1, pp. 21-30.

Argyle, E. (1971). Techniques for edge detection, *Proc. IEEE*, vol. 59, no. 2, pp. 285-287.

Augustson, J. G., and Minker, J. (1970). An analysis of some graph-theoretical cluster techniques, *J. ACM*, vol. 17, no. 4, pp. 571-588.

Avtonomova, V. A. (1973). A statistical model of the observation process in pattern recognition problems, *Autom. Remote Control*, vol. 34, no. 10, Pt. 1, pp. 1566-1574.

Avtonomova, V. A., and Kholodilov, Y. M. (1973). Experimental investigation of the visual image recognition characteristics, *Autom. Remote Control*, vol. 34, no. 8, Pt. 2, pp. 1342-1346.

Babu, C., and Chitti, V. (1973). On the application of probabilistic distance measures for the extraction of feature from imperfectly labeled patterns, *Int. J. Comput. Inf. Sci.*, vol. 2, no. 2, pp. 103-114.

Badhwar, G. D., Lyndon, B., Austin, W. W., and Carnes, J. G. (1982). A semi-automatic technique for multitemporal classification of a given crop within a Landsat scene, *Pattern Recognition*, vol. 15, no. 3 pp. 217-230.

Bakke, K. I., and McMurtry, G. J. (1969). A pattern recognition algorithm using the concept of intrinsic dimensionality, *Proc. Purdue Symp. Inf. Process.*, Apr. 1969, pp. 446-452.

Ball, G. H. (1965). Data analysis in the social sciences: What about the details? *Proc. Fall Joint Comput. Conf.*

Ball, G. H., and Hall, D. J. (1965). Isodata, an iterative method of multivariate analysis and pattern classification, *Proc. IFIPS Congr.*

Ball, G. H., and Hall, D. J. (1965). Isodata, a novel method of data analysis and pattern classification, *NTIS Rep. AD699616*, Tech. Rep. of SRI project, Stanford Research Institute, Menlo Park, Calif.

Bargel, B. (1980). Classification of remote sensed data by texture and shape features in different spectral channels, *Proc. 5th Int. Conf. Pattern Recognition*, Miami Beach, Fla., Dec. 1-4, 1980 (IEEE, New York, 1980), pp. 2-4.

Barnard, S. T. (1980). Automated inspection using grey-scale statistics, *Proc. 5th Int. Conf. Pattern Recognition*, Miami Beach, Fla., Dec. 1-4, 1980, (IEEE, New York, 1980), pp. 269-272.

Barnes, D. I., and Silverman, H. E. (1972). A class of algorithms for fast digital image registration, *IEEE Trans. Comput.*, vol. C-21, no. 2, pp. 179-186.

Batchelor, B. G. (1973). Instability of the decision surfaces of the nearest-neighbor and potential function classifiers, *Inf. Sci.*, vol. 5, p. 179.

Batchelor, B. G. (1978). Experimental and pragmatic approaches to pattern recognition, *Kybernetes*, vol. 7, no. 4, pp. 269-277.

Batchelor, B. G., and Hand, D. J. (1975). On the graphical analysis of PDF estimators for pattern recognition, *Kybernetes*, vol. 4, p. 239.

Batchelor, B. G., and Wilkins, B. R. (1969). Method for location of clusters of patterns to initialize a learning machine, *Electron. Lett.*, vol. 5, no. 20, pp. 481-483.

Bebb, J. E., and Stromberg, W. D. (1973). An overview of image processing, *System Development Corp. Rep. AD-757119.*

Becker, P. W., and Nielson, D. A. (1972). Pattern recognition using dynamic pictorial information, *IEEE Trans. Syst., Man Cybern.*, vol. SMC-2, no. 3, pp. 434-437.

Bell, D. A. (1973). Decision trees in pattern recognition, Ph.D. thesis, University of Southampton.

Benjamin, R. (1980). Generalizations of maximum entropy pattern analysis, *Proc. IEEE*, vol. 27, no. 5, pp. 341-353.

Bennett, J. R., and MacDonald, J. S. (1975). On the measurement of curvature in a quantized environment, *IEEE Trans. Comput.*, vol. C-2, no. 8, pp. 803-820.

Bergland, D. (1969). Fast Fourier transform hardware implementation: a survey, *IEEE Trans. Audio Electroacoust.*, vol. AU-17, pp. 109-119.

Billingsley, F. C. (1970). Applications on digital image processing, *Appl. Opt.*, vol. 2, no. 2, pp. 289-299.

Billingsley, F. C. (1972). Digital image processing for information extraction, in *Machine Perceptions of Pattern and Pictures*, Conf. Ser. 13, Institute of Physics, London, pp. 337-362.

Blaydon, C. C. (1967). Recursive algorithms for pattern classification, *Office Naval Res. Tech. Rep. 520*, Division of Engineering and Applied Physics.

Block, H. D. (1962). The perceptron: a model for brain functioning — I, *Rev. Mod. Phys.*, vol. 34, no. 1, pp. 123-135.

Block, H. D., Nilsson, N. J., and Duda, R. O. (1964). Determination and detection of features in patterns, in *Computer and Information Sciences*, vol. I (J. T. Tou and R. H. Wilcox, eds.), Spartan Books, Washington, D.C.

Blum, J. R. (1954). Approximation methods which converge with probability one, *Ann. Math. Stat.*, vol. 25, pp. 382-386.

Bow, S. T. (1962). Graphical analysis of the non-linear power system oscillation, *Sci. Sin.*, vol. 11, no. 1, 1960, pp. 131-144.

Bow, S. T. (1963). Matrix analysis of pulse wave propagation on multi-conductor system, *Sci. Sin.*, vol. 12, no. 2, pp. 245-270.

Bow, S. T. (1979). Morphological analysis of simplified Chinese ideographs and a heuristic approach for their machine recognition, *Proc. COMPSAC 79*, Chicago, 1979.

Bow, S. T. (1980). Structural approach applicable to primitive description and extraction for complex Chinese ideograph recognition, *Proc. 5th Int. Conf. Pattern Recognition*, Miami Beach, Fla., Dec. 1-4, 1980; IEEE, New York, 1980.

Bow, S. T. (1982). Some considerations on the machine recognition of Chinese ideographs, *Proc. 1982 Int. Conf. Chinese Language Comput. Soc.*, Washington, D.C.

Bow, S. T., and Chu, L. S. (1960). Automatic power system swing curve computing device, *Sci. Rec.*, no. 2, pp. 133-138.

Bow, S. T., and Kim, W. Y. (1982). Preliminary investigation on the structure of Korean characters and their machine recognition, *Proc. 1982 Int. Conf. Chinese Language Comput. Soc.*, Washington, D.C.

Bow, S. T., and Van Ness, J. E. (1958). Use of phase space in transient stability studies, *Trans. AIEE*, Pt. II, p. 77.

Bow, S. T., Wang, C. C., H. Y. Yau, K. N. Tsou (1964). Logic control technique applicable to power system fault diagnosis and handling, *Sci. Sin.*, vol. 13, no. 8, pp. 119-122.

Brakke, K. A., Mantock, J. M., and Fukunaga, K. (1982). Systematic feature extraction, *IEEE Trans. Pattern Anal. Mach. Intell.*, vol. PAMI-4, no. 3, pp. 291-297.

Brauner, R., and Epstein, D. (1981). Automated chip (die) inspection, *Int. J. Hybrid Microelectron*, vol. 4, no. 2, pp. 111-115.

Braveman, E. M. (1965). On the method of potential functions, *Autom. Remote Control*, vol. 26, no. 12, pp. 2130-2138.

Brown, B. R., and Evans, S. H. (1969). Perceptual learning in pattern discrimination tasks with two and three scheme categories, *Psychonom. Sci.*, vol. 15, no. 3, pp 101-103.

Brown, D., Hall, M., and Lal, S. (1970). Pattern transformation by neural nets, *J. Physiol.*, vol. 209, p. 7.

Brown, N., Oreb, B. F., and Hariharen, P. (1982). Pseudocolor picture display with a microcomputer, *J. Phys. E.*, vol. 15, no. 7, pp. 703-704.

Brown, R. (1963). Logical properties of adaptive networks, *Stanford Electron. Lab. Q. Res. Rev.*, no. 4.

Burckhardt, C. B. (1968). Information reduction in holograms for visual displays, *J. Opt. Soc. Am.*, vol. 58, no. 2, pp. 241-246.

Bush, D. A. (1981). Automatic feature extraction system test bed, *IEEE Comput. Soc. Conf. Pattern Recognition Image Process.*, Dallas, Tx., Aug. 3-5, 1981, pp. 615-617.

Butt, E. B. et al. (1968). Studies in visual textual manipulation and synthesis, *Tech. Rep. 68-64*, Computer Science Center, University of Maryland, College Park.

Caulfield, H. J., Haimes, R., and Draper, J. S. (1979). Preprocessing of multispectral images for image processing, *Proc. Soc. Photo-opt. Instrum. Eng.*, vol. 186, pp. 131-134.

Chang, C. L., and Lee, R. C. T. (1973). A heuristic relaxation method for nonlinear mapping in cluster analysis, *IEEE Trans. Syst. Man Cybern.*, vol. SMC-3, no. 2, pp. 197-200.

Chang, S. K. (1971). The reconstruction of binary patterns from their projection, *Commun. ACM*, vol. 14, no. 1, pp. 21-25.

Charnes, A. (1964). On some fundamental theorems of perception theory and their geometry, in *Computer and Information Sciences*, vol. I (J. T. Tou and R. H. Wilcox, eds.), Spartan Books, Washington, D.C.

Chaudhuri, B. B. (1982). Bayes' error and its sensitivity in statistical pattern recognition in noisy environment, *Int. J. Syst. Sci.*, vol. 13, no. 5, pp. 559-570.

Chen, C. H., and Yen, C. (1982). Object isolation in FLIR images using Fischer's linear discriminant, *Pattern Recognition*, vol. 15, no. 3, pp. 153-159.

Chen, P. H., and Wintz, P. A. (1976). Data Compression for satellite images, *TR-EE-76-9*, School of Electrical Engineering, Purdue University, West Lafayette, Ind.

Chien, Y. T. (1970). The threshold effect of a nonlinear learning algorithm for pattern recognition, *Inf. Sci.*, vol. 2, pp. 351-368.

Chien, Y. T. (1976). Interactive pattern recognition: techniques and system, *IEEE Comput.*, vol. C-9, no. 5, May 1, 1976.

Chien, Y. T., and Fu, K. S. (1967). On the generalized Karhunen-Loeve expansion, *IEEE Trans. Inf. Theory*, vol. IT-13, no. 3, pp. 518-520.

Chien, Y. T., and Ribak, R. (1971). Relationship matrix as a multidimensional data base for syntactic pattern generation and recognition, *Proc. Two-Dimensional Signal Process. Conf.*, University of Missouri, Columbia.

Chin, R. T. (1982). Automated visual inspection techniques and applications: a bibliography, *Pattern Recognition*, vol. 15, no. 4, pp. 343-357.

Chittineni, C. B. (1982). Some approaches to optimal cluster labeling with applications to remote sensing, *Pattern Recognition*, vol. 15, no. 3, pp. 201-216.

Chomsky, N. (1956). Three models for the description of language, *POIT*, vol. 2, no. 3, pp. 113-124.

Clark, D. C., and Gonzalez, R. C. (1981). Optimal solution of linear inequalities with applications to pattern recognition, *IEEE Trans. Pattern Anal. Mach. Intell.*, vol. PAMI-3, no. 6, pp. 643-655.

Clowes, M. B. (1969). Transformational grammars and the organization of pictures, in *Automatic Interpretation and Classification of Images* (A. Grasselli, ed.), Academic Press, New York.

Clowes, M. B. (1971). On seeing things, *Artif. Intell.*, vol. 2, pp. 79-116.

Clowes, M. B. (1972). Scene analysis and picture grammars, in *Machine Perception of Pattern and Pictures*, Conf. Ser. 13, Institute of Physics, London, pp. 243-256.

Cofer, R. H. (1972). Picture acquisition and graphical preprocessing system, *Proc. 9th Annu. IEEE Region III Conv.*, Charlottesville, Va.

Cofer, R. H., and Tou, J. T. (1971). Preprocessing for pictorial pattern recognition, *Proc. 21st NATO Tech. Symp. Artif. Intell.*, Italy.

Cofer, R. H., and Tou, J. T. (1972). Automated map reading and analysis by computer, *Proc. Fall Joint Comput. Conf.*

Collins, D. C., and Meisel, W. S. (1971). Structural analysis of biological wave forms, *Proc. Conf. Eng. Med. Biol.*, Las Vegas, Nev., Oct. 31-Nov. 4, 1971.

Colwell, R. N. (1965). The extraction of data from aerial photograph by human and mechanical means, *Photogrammetrics*, vol. 20, pp. 211-228.

Cooley, J. W., and Turkey, J. W. (1965). An algorithm for the machine computation of complex Fourier series, *Math. Comp.*, vol. 19, pp. 297-301.

Cooper, D. R., and Cooper, P. W. (1964). Nonsupervised adaptive
    signal detection and pattern recognition, *Inf. Control*, vol. 7,
    no. 3, pp. 416-444.
Cooper, P. W. (1964). Hyperplanes, hyperspheres and hyperquad-
    rics as decision boundaries, in *Computer and Information
    Sciences*, vol. I (J. T. Tou and R. H. Wilcox, eds.), Spartan
    Books, Washington, D.C.
Cooper, P. W. (1967). Some topics in nonsupervised adaptive detec-
    tion for multivariate normal distributions, in *Computer and In-
    formation Sciences*, vol. II (J. T. Tou, ed.), Academic Press,
    New York.
Cormack, R. M. (1971). A review of classification, *J. R. Stat. Soc.*,
    Ser. A, vol. 134, Pt. E.
Cover, T. M. (1964). Classification and generalization capabilities of
    linear threshold units, *Rome Air Development Center Tech. Rep.
    RADC-TDR-64-32*.
Cover, T. M. (1965). Geometrical and statistical properties of sys-
    tems of linear inequalities with applications to pattern recogni-
    tion, *IEEE Trans. Electron. Comput.*, vol. EC-14, no. 3,
    pp. 326-334.
Cover, T. M. (1969). Learning in pattern recognition, in *Methodolo-
    gies of Pattern Recognition* (S. Watanabe, ed.), Academic Press,
    New York.
Cover, T. M., and Hart, P. E. (1967). Nearest-neighbor pattern
    classification, *IEEE Trans. Inf. Theory*, vol. IT-13, no. 1,
    pp. 21-27.
Crespi-Reghizzi, S. (1971). An effective model for grammar infer-
    ence, *IFIP Congr.* — 71, Yugoslavia.
Dammann, J. E. (1966). An experiment in cluster detection, *IBM J.
    Res. Dev.*, vol. 80, Jan. no. 1.
Daniell, G. J. (1980). Maximum entropy algorithm applied to image
    enhancement, *Proc. IEEE*, vol. 127, no. 5, pp. 170-172.
Davies, D. L., and Bouldin, D. W. (1979). A cluster separation
    measure, *IEEE Trans. Pattern Anal. Mach. Intell.*, vol. PAMI-1,
    no. 2, pp. 224-227.
Davies, L. S. (1980). Computer architectures for image processing,
    Proc. Workshop Picture Data Descr. Manag. (IEEE, New York,
    1980), pp. 249-254.
Dehne, J. S. (1978). Image processing techniques for real-time
    imagery, *Pattern Recognition Signal Process.*, Paris, June 25-
    July 4, 1978.
Denes, P. B. (1975). A scan type graphics system for interactive
    computing, *Proc. Conf. Comput. Graphics, Pattern Recognition
    Data Struc.*, May 1975 (IEEE Cat. 75CHO 981-1C), pp. 21-24.
DeSouza, P. (1982). Some decision network designs for pattern class-
    ification, *Pattern Recognition*, vol. 15, no. 3, pp. 193-200.

Diday, E. (1973). The dynamic clusters method in non-hierarchical clustering, *Int. J. Comput. Inf. Sci.*, vol. 2, no. 1, pp. 61-88.

Diday, E. (1980). Clustering in pattern recognition, *Proc. 5th Int. Conf. Pattern Recognition*, Miami Beach, Fla., Dec. 1-4, 1980 (IEEE, New York, 1980), pp. 424-429.

Diday, E., and Celeux, G. (1981). Optimization in cluster analysis, *Methods Oper. Res.*, no. 43, p. 327.

Duan, J. R., and Wintz, P. A. Information preserving coding for multispectral data, *Proc. IEEE Conf. Mach. Process. Remotely Sensed Data.*

Duan, J. R., and Wintz, P. A. (1974). Information preserving coding for multispectral scanner data, *TR-EE-74-15*, School of Electrical Engineering, Purdue University, West Lafayette, Ind.

Duda, R. O., and Fossum, H. (1966). Pattern classification by iteratively determined linear and piecewise linear discriminant functions, *IEEE Trans. Electron. Comput.*, vol. EC-15, no. 2, pp. 220-232.

Duff, M. J. B., Norgren, P., Preston, J., Jr., and Toriwaki (1978). Theoretical and practical consideration in the application of neighborhood logic to image processing, *Proc. 4th Int. Joint Conf. Pattern Recognition*, Kyoto, Japan, Nov. 7-10, 1978, pp. 139-145.

Duff, M. J. B., Watson, D. M., and Deutsch, E. S. (1974). A parallel computer for array processing, *Inf. Process. — 74 (Proc. IFIP Congr.)*, pp. 94-97.

Duff, M. J. B., Watson, D. M., Fountain, T. J., and Shaw, G. K. (1973). A cellular logic array for image processing, *Pattern Recognition*, vol. 5, pp. 229-247.

Dvoretzky, A. (1956). On stochastic approximation, in *Proceedings of the Third Berkely Symposium on Mathematical Statistics and Probability* (J. Neyman, ed.), University of California Press, Berkeley, pp. 39-55.

Dyer, C. R., and Rosenfeld, A. (1981). Parallel image processing by memory-augmented cellular automata, *IEEE Trans. Pattern Anal. Mach. Intell.*, vol. PAMI-3, no. 1, pp. 29-41.

Edwards, J. A., and Fitelson, M. M. (1973). Notes on maximum entropy processing, *IEEE Trans. Inf. Theory*, vol. IT-19, no. 2, pp. 232-234.

Ehrich, R. W. (1978). A view of texture topology and texture description, *Comput. Graphics Image Process.*, vol. 8, no. 2, pp. 174-202.

Eklundh, J. O. (1972). A fast computer method for matrix transposing, *IEEE Trans. Comput.*, vol. C-21, July, pp. 801-803.

Eklundh, J. O., and Rosenfeld, A. (1981). Image smoothing bases based on neighbor linking, *IEEE Trans. Pattern Anal. Mach. Intell.*, vol. PAMI-3, no. 6, pp. 679-683.

Evans, T. G. (1971). Grammatical inference techniques in pattern analysis, in *Software Engineering* (J. T. Tou, ed.), Academic Press, New York.

Fainzil'berg, L. S. (1978). Pattern recognition methods in thermographic metal composition analysis, *Cybernetics*, vol. 14, no. 6, pp. 951-955.

Farrow, A. S. J. (1974). TV scanner to computer in real time, in *Oxford Conference on Computer Scanning*, vol. 2 (P. G. Davey and B. M. Hawes, eds.), Nuclear Physics Laboratory, Oxford, pp. 407-422.

Feldman, J. (1967). First thoughts on grammatical inference, *Artif. Intell. Memo 55*, Computer Science Dept., Stanford University, Stanford, Calif.

Feldman, J. (1969). Some decidability results on grammatical inference and complexity, *Artif. Intell. Memo 93*, Computer Science Dept., Stanford University, Stanford, Calif.

Feldman, J., Gips, J., and Reder, S. (1969). Grammatical complexity and inference, *Artif. Intell. Memo 89*, Computer Science Dept., Stanford, Calif.

Feldman, J. A., and Yakimovsky, Y. (1975). Decision theory and artificial intelligence: a semantics-based region analyzer, *Artif. Intell.*, vol. 5, no. 4, pp. 349-372.

Feng, H. Y. F., and Pavlidis, T. (1975). Decomposition of polygons into simpler components: feature generation for syntactic pattern recognition, *IEEE Trans. Comput.*, vol. C-24, no. 6, pp. 636-650.

Fenker, R. M., Jr., and Evans, S. H. (1971). A model for optimizing the effectiveness of man-machine decision making in a pattern recognition system, *Tech. Rep.*, Texas Christian University, Institute for the Study of Cognitive Systems, Fort Worth, June 1971, AD-730944.

Findler, N. V. (1979). A family of similarity measures between two strings, *IEEE Trans. Pattern Anal. Mach. Intell.*, vol. PAMI-1, no. 1, pp. 116-118.

Fink, W. (1976). Image coloration as an interpretation aid, *Proc. SPIE/OSA Conf. Image Process.*, Pacific Grove, Calif., vol. 74, pp. 209-215.

Firschein, O., Eppler, W., and Fischler, M. A. (1979). A fast defect measurement algorithm and its array processor mechanization, *Proc. 1979 IEEE Comput. Soc. Conf. Pattern Recognition Image Process.*, pp. 109-113.

Firschein, O., Rauch, H. E., and Eppler, W. G. (1980). Track assembly and background suppression using an array processor and neighborhood coding, *Proc. Soc. Photo-opt. Instrum. Eng.*, vol. 241, pp. 258-266.

Fischler, M. A. (1969). Machine perception and description of pictorial data, *Proc. Joint Int. Conf. Artif. Intell.*, Washington, D.C.

Fisk, C. J. (1972). Imagine a computer program for image enhancement and display of two-dimensional data, *Sandia Laboratories Res. Rep. SC-RR-72-0286*, Albuquerque, N. Mex., July 1972.

Fitts, J. M. (1980). Automatic target identification of blurred images with superresolution features, *Proc. Soc. Photo-opt. Instrum. Eng.*, vol. 252, pp. 85-91.

Fix, E., and Hodges, J. L., Jr. (1951). Discriminatory analysis, nonparametric discrimination, *Project 21-49-004, Rep. 4*, U.S. Air Force School of Aviation Medicine, Randolph Field, Tex. (Contract AF 41-128-31).

Foley, D. H. (1971). The probability of error on the design set as a function of the sample size and dimensionality, *Tech. Rep., RACD-TR-71-171*, Griffiss Air Force Base, N.Y., Dec. 1971.

Foley, D. H. (1972). Considerations of sample and feature,size, *IEEE Trans. Inf. Theory*, vol. IT-18, no. 5, pp. 618-626.

Foley, D. H., and Sammon, J. W. (1975). An optimal set of discriminant vectors, *IEEE Trans. Comput.*, vol. C-24, no. 3, pp. 281-289.

Ford, N. L., Batchelor, B. G., and Wilkins, B. R. (1970). Learning scheme for the nearest neighbor classifier, *Inf. Sci.*, vol. 2, p. 139.

Freeman, H. (1977). Computer processing of line drawing images, *ACM Comput. Surv.*, vol. 6, pp. 57-97.

Frei, W., and Chen, C. C. (1977). Fast boundary detection: a generalization and a new algorithm, *IEEE Trans. Comput.*, vol. TC-26, pp. 988-998.

Frieden, B. R. (1972). Restoring with maximum likelihood and maximum entropy, *J. Opt. Soc. Am.*, vol. 62, pp. 511-518.

Frieden, B. R. (1975). Image enhancement and restoration, in *Picture Processing and Digital Filtering* (T. S. Huang, ed.), Springer-Verlag, Berlin, pp. 179-246.

Frieden, B. R., and Burke, J. J. (1972). Restoring with maximum entropy — II. Superresolution of photographs of diffraction blurred images, *J. Opt. Soc. Am.*, vol. 62, pp. 1207-1210.

Fu, K. S. (1970). Stochastic automata as models of learning systems, in *Adaptive, Learning, and Pattern Recognition Systems* (J. M. Mendel and K. S. Fu, eds.), Academic Press, New York.

Fu, K. S. (1971). On syntactic pattern recognition and stochastic languages, *Tech. Rep. TR-EE-71-21*, School of Electrical Engineering, Purdue University, Lafayette, Ind.

Fu, K. S. (1972). A survey of grammatical inference, *Tech. Rep. TR-EE-72-13*, School of Electrical Engineering, Purdue University, Lafayette, Ind.

Fu, K. S. (1978). Recent advances in syntactic pattern recognition, *Proc. 4th Int. Joint Conf. Pattern Recognition*, Kyoto, Japan, Nov. 7-10, 1978.

Fu, K. S. (1979). Size normalization and pattern orientation problems in syntactic clustering, *IEEE Trans. Syst. Man Cybern.*, vol. SMC-9, no. 1, pp. 55-58.

Fu, K. S. (1980). Recent developments in pattern recognition, *IEEE Trans. Comput.*, vol. C-29, no. 10, pp. 845-856.

Fu, K. S., and Bhargava, B. K. (1973). Tree systems for syntactic pattern recognition, *IEEE Trans. Comput.*, vol. C-22, no. 12, pp. 1087-1099.

Fu, K. S., and Swain, P. H. (1971). On syntactic pattern recognition, in *Software Engineering* (J. T. Tou, ed.), Academic Press, New York.

Fukunaga, K. (1981). The optimal distance measure for nearest neighbor classification, *IEEE Trans. Inf. Theory*, vol. IT-27, no. 5, pp. 622-627.

Fukunaga, K., and Koontz, W. L. C. (1970). Application of the Karhunen-Loeve expansion to feature selection and ordering, *IEEE Trans. Comput.*, vol. C-19, no. 4, pp. 311-318.

Fukunaga, K., and Olsen, D. R. (1971). An algorithm for finding intrinsic dimensionality of data, *IEEE Trans. Comput.*, vol. C-20, pp. 176-183.

Fukunaga, K., and Mantock, J. M. (1982). A non-parametric two-dimensional display for classification, *IEEE Trans. Pattern Anal. Mach. Intell.*, vol. PAMI-4, no. 4, pp. 427-436.

Galloway, M. M. (1975). Texture analysis using grey level run lengths, *Comput. Graphics Image Process.*, vol. 4, no. 2, pp. 172-179.

Gilbert, A. L., Giles, M. K., Flacks, G. M., and Rogers, R. B. (1979). A real time video tracking system, *Opt. Eng.*, vol. 18, no. 1, pp. 25-32. Also *Proc. 4th Int. Joint Conf. Pattern Recognition*, Kyoto, Japan, Nov. 7-10, 1978, p. 111015; *IEEE Trans. Pattern Anal. Mach. Intell.*, vol. PAMI-2, no. 1, pp. 47-56 (1980).

Gillenson, M. L., and Chandrasikaran, B. A. (1975). A neuristic strategy for developing images on a CRT, *Pattern Recognition*, vol. 7, no. 4, pp. 187-196.

Gold, E. M. (1967). Language identification in the limit, *Inf. Control*, vol. 10, no. 5, pp. 447-474.

Gonzalez, R. C. (1972). Syntactic pattern recognition — introduction and survey, *Proc. Nat. Electron. Conf.*, vol. 27, no. 1, pp. 27-32.

Gonzalez, R. C. (1973). Generation of linguistic filter structures for image enhancement, *Proc. ACM Conf.*

Gonzalez, R. C., and Fittes, B. A. (1975). Grey-level transformations for interactive image enhancement, *Mech. Mach. Theory*, vol. 12, pp. 111-112.

Gonzalez, R. C., Fry, D. N., and Kryter, R. C. (1974). Results in the application of pattern recognition methods to nuclear reactor core component surveillance, *IEEE Trans. Nucl. Sci.*, vol. 21, no. 1, pp. 750-757.

Gonzalez, R. C., Lane, M. C., Bishop, A. O., Jr., and Wilson, W. P. (1972). Some results in automatic sleep-state classification, *Proc. 4th Southeastern Symp. Syst. Theory*.

Gonzalez, R. C., and Thomason, M. G. (1974). Tree grammars and their application to pattern recognition, *Tech. Rep. TR-EE-CS-74-10*, Electrical Engineering Dept., University of Tennessee, Knoxville.

Gonzalez, R. C., and Thomason, M. G. (1974). Inference of tree grammars for syntactic pattern recognition, *Tech. Rep. TR-EE/CS-74-20*, Electrical Engineering Dept., University of Tennessee, Knoxville.

Gonzalez, R. C., and Tou, J. T. (1968). Some results in minimum entropy feature extraction, *IEEE Conv. Rec. — Region III*.

Gonzalez, R. C., and Wagner, C. G. (1982). Moments and semi-invariants of the interclass Mahalanobis distance, *Proc. PRIP '82*, pp. 12-17.

Granlund, G. H. (1978). In search of a general picture processing operator, *Comput. Graphics Image Process.*, vol. 8, no. 2, pp. 153-173.

Gray, S. B. (1971). Local properties of binary images in two dimensions, *IEEE Trans. Comput.*, vol. C-20, no. 5, pp. 551-561.

Greblicki, W. (1978). Pattern recognition procedures with non-parametric density estimates, *IEEE Trans. Syst. Man. Cybern.*, vol. SMC-8, no. 11, pp. 809-812.

Guerra, C., and Pieroni, G. G. (1982). A graph-theoretic method for decomposing two-dimensional polygonal shapes into meaningful parts, *IEEE Trans. Pattern Anal. Mach. Intell.*, vol. PAMI-4, no. 4, pp. 405-408.

Gunther, F. J. (1982). A new principal components procedure to aid the analysis of Landsat MSS digital data, *Proc. PRIP '82*, pp. 38-43.

Guzman, A. (1968). Decomposition of a visual scene into three-dimensional bodies, *Proc. Fall Joint Comput. Conf.*

Hall, E. L. (1974). A comparison of computations for spatial frequency filtering, *Proc. IEEE*, vol. 60, no. 7, pp. 887-891.

Hall, E. L. (1974). Almost uniform distribution for computer image enhancement, *IEEE Trans. Comput.*, vol. C-23, no. 2, pp. 207-208.

Hall, E. L., Kruger, R. P., Dwyer, S. L., Hall, D. L., McLaren, R. W., and Lodwick, G. S. (1971). A survey of preprocessing and feature extraction techniques for radiographic images, *IEEE Trans. Comput.*, vol. C-20, pp. 1032-1044.

Hall, E. L., Tie, J. B. K., McPherson, C. A., and Sadjadi, F. A. (1982). Curved surface measurement and recognition for robot vision, *Conf. Rec. 1982 Workshop Ind. Appl. Mach. Vision*, Research Triangle Park, N.C., May 3-5, 1982, pp. 187-199.

Hanaizumi, H., Inamura, M., Toyota, H., and Fujmura, . (1980). Noise reduction of multi-spectral scanner image data using scan overlap, *Trans. Soc. Instrum. Control Eng. (Japan)*, vol. 16, no. 6, pp. 880-885.

Hanakata, K. (1974). A descriptive data structure for interactive pattern recognition, *Proc. 2nd Int. Joint Conf. Pattern Recognition*, Copenhagen, Aug. 13-15, 1974, pp. 421-423.

Hand, D. J., and Batchelor, B. G. (1975). Classification of incomplete pattern vectors using orthogonal function methods, *Proc. 3rd Int. Conf. Cybern. Syst.*, Bucharest.

Hankley, W. J., and Tou, J. T. (1968). Automatic fingerprint interpretation and classification via contextual analysis and topological coding, in *Pictorial Pattern Recognition* (G. C. Cheng et al., eds.), Thompson Books, Washington, D.C.

Hannigan, J. F., and Gerhart, G. R. (1980). Direct electronic Fourier transform (DEFT) spectra for surveillance and countersurveillance, *Proc. Soc. Photo-opt. Instrum. Eng.*, vol. 241, pp. 113-121.

Haralick, R. M. (1971). Multi-image clustering, *Proc. 1970 Army Numer. Analy. Conf.*, U.S. Army Research Office, Rep. 71-1, Durham, N.C., Jan. 1971, pp. 75-90.

Haralick, R. M. (1978). Statistical and structural approaches to texture, *Proc. 4th Int. Joint Conf. Pattern Recognition*, Kyoto, Japan, Nov. 7-10, 1978, pp. 45-69.

Haralick, R. M., and Kelley, G. L. (1969). Pattern recognition with measurement space and spatial clustering for multiple images, *Proc. IEEE*, vol. 57, no. 4, pp. 654-665.

Haralick, R. M., and Shapiro, L. G. (1979). Decomposition of two-dimensional shapes by graph-theoretic clustering, *IEEE Trans. Pattern Anal. Mach. Intell.*, vol. PAMI-1, no. ±, pp. 10-20.

Hartley, R. L., Kitchen, L. J., Wang, C. Y., and Rosenfeld, A. (1982). Segmentation of FLIR images: a comparative study, *IEEE Trans. Syst. Man Cybern.*, vol. SMC-12, no. 4, pp. 553-566.

Hawkins, J. K. (1970). Image processing principles and techniques, in *Advances in Information System Science* (J. T. Tou, ed.), vol. 3, Plenum Press, New York.

Heintzen, P. H., Malerczyk, H., and Scheel, K. W. (1971). On-line processing of videoimages for left ventrical volume determination, *Comput. Biomed. Res.*, vol. 4, no. 5.

Heller, J. (1978). Image motion restoration, *Tech. Rep.*, Dept. of Electrical Engineering, University of Tennessee, Knoxville, May.

Herbst, N. W., and Will, P. M. (1972). An experimental laboratory for pattern recognition and signal processing, *Commun. ACM*, vol. 15, no. 4, pp. 231-244.

Herman, G. T. (1972). Two direct methods for reconstructing pictures from their projections — a comparative study, *Comput. Graphics Image Process.*, vol. 1, no. 2, pp. 123-144.

Highleyman, W. H. (1961). Linear decision functions with applications to pattern recognition, Ph.D. dissertation, Electrical Engineering Dept., Polytechnic Institute of Brooklyn, New York. [A summary bearing the same title may be found in *Proc. IRE*, vol. 50, no. 6, pp. 1501-1514 (1962).]

Ho, Y. C., and Agrawala, A. K. (1968). On pattern classification algorithms — introduction and survey, *Proc. IEEE*, vol. 56, no. 12, pp. 2101-2114.

Ho, Y. C., and Kashyap, R. (1965). An algorithm form linear inequalities and its applications, *IEEE Trans. Electron. Comput.*, vol. EC-14, no. 5, pp. 683-688.

Holdermann, F., and Kazmierczak, H. (1972). Preprocessing of greyscale pictures, *Comput. Graphics Image Process.*, no. 1, pp. 66-80.

Horning, J. J. (1969). A study of grammatical inference, *Tech. Rep. CS-139*, Computer Science Dept., Stanford University, Stanford, Calif.

Horning, J. J. (1971). A procedure for grammatical inference, *IFIP Congr.* — 71, Yugoslavia.

Hotelling, H. (1933). Analysis of a complex of statistical variables into principal components, *J. Educ. Psychol.*, vol. 24, pp. 417-441, 498-520.

Huang, T. S. (1979). Trends in digital image processing research, in *Advances in Digital Image Processing* (Bad Neuenahr, Germany, Sept. 26-28, 1978), Plenum, New York, 1979, pp. 21-30.

Huang, T. S., Schreiber, W. F., and Tretiak, O. J. (1971). Image processing, *Proc. IEEE*, vol. 59, pp. 1586-1609.

Hueckel, M. H. (1973). A local visual operator which recognizes edges and lines, *J. ACM*, vol. 20, pp. 634-647.

Hummel, R. A., and Rosenfeld, A. (1978). Relaxation processes for scene labeling, *IEEE Trans. Syst. Man Cybern.*, vol. SMC-8, no. 10, pp. 765-768.

Hunt, B. R. (1972). Data structures and computational organization in digital image enhancement, *Proc. IEEE*, vol. 60, no. 7, pp. 884-887.

Hunt, B. R. (1975). Digital image processing, *Proc. IEEE*, vol. 63, no. 4, pp. 693-708.

Hunt, B. R., and Sera, G. F. (1978). Power-law stimulus response models for measures of image quality in non-performance environments, *IEEE Trans. Syst. Man Cybern.*, vol. SMC-8, no. 11, pp. 781-791.

Husson, T. R., and Abdalla, A. M. (1981). Real-time infrared image processing, *IEEE Comput. Soc. Conf. Pattern Recognition Image Process.*, Dallas, Aug. 3-5, 1981, pp. 478-480.

Ichino, M. (1981). Nonparametric feature selection method based on local interclass structure, *IEEE Trans. on Syst. Man Cybern.*, vol. SMC-11, no. 4, pp. 289-296.

Inigo, R. M., and McVey, E. S. (1981). CCD implementation of a three dimensional video-tracking algorithm, *IEEE Trans. Pattern Anal. Mach. Intell.*, vol. PAMI-33, no. 2, pp. 230-240.

Inoue, K. (1979). Image processing and pattern recognition in welding engineering, *Syst. Control (Japan)*, vol. 23, no. 7, pp. 370-378 (in Japanese).

Jacks, E. L. (1964). A laboratory for the study of graphical man-machine communication, *AFIPS Conf. Proc.*, vol. 26-1, pp. 343-350.

Jackson, P. H. (1982). Image contrast enhancement by local maximum and local minimum operations, *IBM Tech. Disclosure Bull.*, vol. 24, no. 12, May 1982.

Jacob, R. J. K. (1976). The face as a data display, *Human Factors*, vol. 18, no. 2, pp. 189-200.

Jagadeeson, M. (1970). N-dimensional fast fourier transform, *Proc. 13th Midwest Symp. Circuit Theory*, III, vol. 2, pp. 1-8.

Jain, A. K. (1974). A fast Karhunen-Loeve transform for finite discrete images, *Proc. Nat. Electron. Conf.*, Chicago, Oct. 1974, pp. 323-328.

Jain, A. K., and Angel, E. (1974). Image restoration, modeling and reduction of dimensionality, *IEEE Trans. Comput.*, vol. C-23, pp. 470-476.

Jain, A. K., and Walter, W. G. (1978). On the optimal number of features in the classification of multivariate Gaussian data, *Pattern Recognition*, vol. 10, no. 5-6, pp. 365-374.

Jardine, N., and Sibson, R. (1968). The construction of hierarchical and non-hierarchic classification, *Comput. J.*, vol. 11, pp. 177-184.

Jarvis, R. (1974). An interactive minicomputer laboratory for graphics image processing and pattern recognition, *Computer*, vol. 7, no. 7, pp. 49-60.

Joseph, R. D. (1960). Contributions to perception theory, *Cornell Aeronaut. Lab. Rep. VG-1196-G-7*.

Kabuka, M., and McVey, E. S. (1982). A position sensing method using images, *Proc. 14th Southeast. Symp. Syst. Theory*, Blacksburg, Va., Apr. 15-16, 1982, pp. 191-194.

Kanal, L. N., and Kumar, V. (1981). Parallel implementation of structural analysis algorithm, *IEEE Comput. Soc. Conf. Pattern Recognition Image Process.*, Dallas, Aug. 3-5, 1981, pp. 452-458.

Kanal, L. N., and Randall, N. C. (1964). Recognition system design by statistical analysis, *Proc. 19th ACM Nat. Conf.*

Karhunen, K. (1947). Über lineare Methoden in der Wahrscheinlichkeitsrechnung, *Ann. Acad. Sci. Fenn.*, Ser. A137 (translated by I. Selin in *"On Linear Methods in Probability Theory"*, T-131, The RAND Corp., Santa Monica, Calif., 1960).

Kashyap, R. L., and Chellappa, R. (1981). Stochastic models for closed boundary analysis: representation and reconstruction, *IEEE Trans. Inf. Theory*, vol. IT-27, no. 5, pp. 627-637.

Kashyap, R. L., and Mittal, M. C. (1973). Picture reconstruction from projections, *Proc. First Int. Joint Conf. Pattern Recognition*, Washington, D.C., IEEE Cat. 73CH0 821-9C, pp. 286-292.

Kazmierczak, H. (1973). Problems in automatic pattern recognition, *Proc. Int. Comput. Symp.*, Davos, Switzerland, pp. 357-370.

Keeha, D. G. (1965). A note on learning for Gaussian properties, *IEEE Trans. Inf. Theory*, vol. IT-II, no. 1, pp. 126-132.

Kendall, M. G. (1973). The basic problems of cluster analysis, in *Discriminant Analysis and Applications* (T. Cacoullos, ed.), Academic Press, New York, pp. 179-191.

Ketcham, D. J. (1976). Real time image enhancement technique, *Proc. SPIE/OSA Conf. Image Process.*, Pacific Grove, Calif., vol. 74, Feb. 1976, pp. 120-125.

K. A. Kirsch (1964). Computer interpretation of English text and picture patterns, *IEEE Trans. Electron. Comput.*, vol. EC-13, no. 4, pp. 363-376.

Kitter, J., and Young, P. C. (1973). A new approach to feature selection based on Karhunen-Loeve expansion, *Pattern Recognition*, vol. 5, pp. 335-352.

Kittler, J. (1978). A method for determining k-nearest neighbors, *Kybernetes*, vol. 7, no. 4, pp. 313-315.

Kobayashi, H., and Bahl, L. R. (1974). Image data compression by predictive coding [2 parts], *IBM J. Res. Dev.*, pp. 164-179.

Koford, J. (1962). Adaptive network organization, *Stanford Electron. Lab. Q. Res. Rep. 3*, p. III-6.

Koontz, W. L. G., and Fukunaga, K. (1972). A non-linear feature extraction algorithm using distance transformation, *IEEE Trans. Comput.*, vol. C-21, no. 1, pp. 56-63.

Koontz, W. L. G., and Fukunaga, K. (1972). A non-parametric valley seeking technique for cluster analysis, *IEEE Trans. Comput.*, vol. C-21, no. 2, pp. 171-178.

Kovelevsky, V. A. (1970). Pattern recognition, heuristics or science? in *Advances in Information Systems Science* (J. T. Tou, ed.), vol. 3, Plenum Press, New York.

Kovalevsky, V. A. (1978). Recent advances in statistical pattern recognition, *Proc. 4th Int. Joint Conf. Pattern Recognition*, Kyoto, Japan, Nov. 7-10, 1978, pp. 1-12.

Kruse, B. (1973). A parallel picture processing machine, *IEEE Trans. Comput.*, vol. C-22, no. 12, pp. 1075-1087.

Laboratory for Agricultural Remote Sensing, Annual Report, vol. 4, Agricultural Experiment Station, *Res. Bull. 873*, Dec. 1970, Purdue University, Lafayette, Ind.

Lamar, J. V., Stratton, R. H., and Simac, J. J. (1972). Computer techniques for pseudocolor image enhancement, *Proc. First USA-Japan Comput. Conf.*, pp. 316-319.

Landau, H. J., and Slepian, D. (1971). Some computer experiments in picture processing for bandwidth reduction, *Bell Syst. Tech. J.*, vol. 50, pp. 1525-1540.

Leboucher, G., and Lowitz, G. E. (1979). What a histogram can tell the classifier, *Pattern Recognition*, vol. 10, no. 5-6, pp. 351-357.

Ledley, R. S. (1964). High speed automatic analysis of biomedical pictures, *Science*, vol. 146, no. 3641, pp. 216-223.

Ledley, R. S., et al. (1965). FIDAC: film input to digital automatic computer and associated syntax-directed pattern recognition programming system, in *Optical and Electro-optical Information Processing Systems* (J. Tippet, D. Beckowitz, L. Clapp, C. Koester, and A. Vanderburgh, Jr., eds.), MIT Press, Cambridge, Mass., Chapt. 3.

Lee, H. C., and Fu, K. S. (1971). A stochastic syntax analysis procedure and its application to pattern classification, *Proc. Two-Dimensional Digital Signal Process. Conf.*, University of Missouri, Columbia.

Lee, H. C., and Fu, K. S. (1974). A syntactic pattern recognition system with learning capability, in *Information Systems: COINS IV* (J. T. Tou, ed.), Plenum Press, New York.

Lee, J. S. (1980). Digital image enhancement and noise filtering by use of local statistics, *IEEE Trans. Pattern Anal. Mach. Intell.*, vol. PAMI-2, no. 2, pp. 165-168.

Lee, R. C. T. (1974). Sub-minimal spanning tree approach for large data clustering, *Proc. 2nd Int. Joint Conf. Pattern Recognition*, Copenhagen, p. 22.

Lee, R. C. T. (1981). Clustering analysis and its applications, in *Advances in Information Systems Science* (J. T. Tou, ed.), vol. 8, Plenum Press, New York, pp. 169-292.

Leese, J. A., Novak, C. S., and Taylor, V. R. (1970). The determination of cloud pattern motions from geosynchronous satellite image data, *Pattern Recognition*, vol. 2, pp. 279-292.

Lendaris, G. G., and Stanley, G. L. (1970). Diffraction pattern sampling for automatic pattern recognition, *Proc. IEEE*, vol. 58, pp. 198-216.

Levialdi, S. (1968). CLOPAN: a closed-pattern analyzer, *Proc. IEEE*, vol. 115, pp. 879-880.

Levialdi, S. (1970). Parallel counting of binary patterns, *Electron. Lett.*, vol. 6, pp. 798-800.

Li, C. C., Ameling, W., DeMori, R., Fu, K. S., Harlow, C. A. Hütting, M. K., Pavlidis, T., Pöppl, S. J., Van Bemmel, E. H., and Wood, E. H. (1979). *Cardio-Pulmonary Systems Group Report*, Dahlem Workshop Report on "Biomedical Pattern Recognition and Image Processing", held in Berlin in May 1979, pp. 299-330.

Licklider, J. C. R. (1969). A picture is worth a thousand words and its costs, *AFIPS Conf. Proc.*, vol. 34, pp. 617-622.

Lillestrand, R. L. K. (1972). Techniques for change detection, *IEEE Trans. Comput.*, vol. C-21, no. 7, pp. 654-659.

Lo, C. M. (1980). A survey of FLIR image enhancement, *Proc. 5th Int. Conf. Pattern Recognition*, Miami Beach, Fla., Dec. 1-4, 1980 (IEEE, New York, 1980), pp. 920-924.

Lu, S. Y. (1979). A tree-to-tree distance and its application to cluster analysis, *IEEE Trans. Pattern Anal. Mach. Intell.*, vol. PAMI-1, no. 2, pp. 219-224.

Lu, S. Y., and Fu, K. S. (1978). Error-correcting tree automata for syntactic pattern recognition, *IEEE Trans. Comput.*, vol. C-27, no. 11, pp. 1040-1053.

Lucas, B. T., and Gardner, K. L. (1980). A generalized classification technique, *Proc. 5th Int. Conf. Pattern Recognition*, pp. 647-653.

Lundsteen, C., Gerdes, T., and Phillip, K. (1982). A model for selection of attributes for automatic pattern recognition — stepwise data compression monitored by visual classification, *Pattern Recognition*, vol. 15, no. 3, pp. 243-251.

Lutz, R. K. (1980). An algorithm for the real-time analysis of digitized images, *Comput. J.*, vol. 23, no. 3, pp. 262-269.

Machine Processing for Remotely Sensed Data, *Conf. Proc.*, Purdue University, Oct. 1973, IEEE Cat. 73CH0 834-2GE.

Matsushita, Y. (1972). Hidden lines elimination for a rotating object, *Commun. ACM*, vol. 15, no. 4, pp. 245-252.

McCormick, B. H. (1963). The Illinois pattern recognition computer — ILLIAC III, *Trans. IEEE Electron. Comput.*, vol. EC-12, pp. 792-813.

McMurtry, G. J. (1976). Preprocessing and feature selection in pattern recognition with application to remote sensing and multispectral scanner data, *IEEE 1976 Int. Conf. on Cybern.*, Washington, D.C., Nov. 1-3, 1976.

Merill, T., and Green, D. M. (1963). On the effectiveness of receptors in recognition systems, *IEEE Trans. Inf. Theory*, vol. IT-9, no. 1, pp. 11-27.

Mero, L. (1980). Edge extraction and line following using parallel processing, *Proc. Workshop Picture Data Descr. Manag.*, IEEE, New York, 1980, pp. 255-258.

Miller, W. F., and Linggard, R. (1982). A perceptually based spectral distance measure, *1982 Int. Zurich Semin. Digital Commun. Man-Mach. Interact.*, Zurich, Mar. 9-11, 1982, E4/143-6.

Miller, W. F., and Shaw, A. C. (1968). Linguistic methods in picture processing — a survey, *Proc. Fall Joint Comput. Conf.*

Minsky, M. L. (1961). Steps toward artificial intelligence, *Proc. IRE*, vol. 49, no. 1, pp. 8-30.

Minter, T. C., Lennington, R. K., and Chittineni, C. B. (1981). Probabilistic cluster labeling of imagery data, *IEEE Comput. Soc. Conf. Pattern Recognition Image Process.*, 1981.

Mizoguchi, R. and Kakusho, O. (1978). Hierarchical clustering algorithm based on k-nearest neighbors, *4th Int. Joint Conf. on Pattern Recognition*, Kyoto, Japan, pp. 314-316.

Mori, R. I., and Raeside, D. E. (1981). A reappraisal of distance-weighted k-nearest neighbor classification for pattern recognition with missing data, *IEEE Trans. Syst. Man Cybern.*, vol. SMC-11, no. 3, pp. 241-243.

Mott-Smith, J. C., Cook, F. H., and Knight, J. M. (1972). Computer aided evaluation of reconnaissance image compression schemes using an on-line interactive facility, *Phys. Sci. Res. Paper 480*, AFCRL-72-0115, Feb. 1972.

Mucciardi, A. N., and Gose, E. E. (1972). An automatic clustering algorithm and its properties in high dimensional spaces, *Trans. IEEE Syst. Man. Cybern.*, vol. SMC-2, p. 247.

Mucciardi, A. N., and Gose, E. E. (1972). Comparison of seven techniques for choosing subsets of pattern recognition properties, *Trans. IEEE Comput.*, vol. C-20, p. 1023.

Mullin, J. K. (1982). Interfacing criteria for recognition logic used with a context post-processor, *Pattern Recognition*, vol. 15, no. 3, pp. 271-273.

Murray, G. G. (1972). Modified transforms in imagery analysis, *Proc. 1972 Symp. Appl. Walsh Functions*, pp. 235-239.

Nagao, S. M., and Fukunaga, Y. (1974). An interactive picture processing system on a minicomputer, *Proc. 2nd Int. Conf. Pattern Recognition*, Copenhagen, Aug. 13-15, 1974, pp. 148-149.

Nagy, G. (1968). State of the art in pattern recognition, *Proc. IEEE*, vol. 56, no. 5, pp. 836-862.

Nahi, N. E. (1972). Role of recursive estimation in statistical image enhancement, *Proc. IEEE*, vol. 60, pp. 872-877.

Narasimhan, R. (1962). A linguistic approach to pattern recognition, *Rep. 21*, Digital Computer Laboratory, University of Illinois, Urbana.

Narayanan, K. A., and Rosenfeld, A. (1981). Image smoothing by local use of global information, *IEEE Trans. Syst. Man Cybern.*, vol. SMC-11, no. 12, pp. 826-831.

Navarro, A. (1976). The role of the associative processor in pattern recognition, *Proc. NATO Advanced Studies Inst.*, Bandol, France, Sept. 1975.

Nitzan, D., and Agin, G. J. (1979). Fast methods for finding object outlines, *Comput. Graphics Image Process.*, vol. 9, no. 1, pp. 22-39.

O'Handley, D. A., and Green, W. B. (1972). Recent development in digital image processing at the image processing laboratory at the Jet Propulsion Laboratory, *Proc. IEEE*, vol. 60, no. 7, pp. 821-828.

Osteen, R. E., and Tou, J. T. (1973). A clique detection algorithm based on neighborhoods in graphs, *Int. J. Comput. Inf. Sci.*, vol. 2, no. 4, pp. 257-268.

Otsu, N. (1979). A threshold selection method from grey level histograms, *IEEE Trans. Syst. Man Cybern.*, vol. SMC-9, no. 1, pp. 62-66.

Pal, S. K. (1982). Optimum guard zone for self supervised learning, *Proc. IEEE*, vol. 129, no. 1, pp. 9-14.

Panda, D., Aggarwal, R., and Hummel, R. (1980). Smart sensors for terminal homing, *Proc. Soc. Photo-opt. Instrum. Eng.*, vol. 252, pp. 94-97.

Pao, T. W. (1969). A solution of the syntactic induction inference problem for a non-trivial subset of context free languages, *Interim Tech. Rep. 78-19*, Moore School of Electrical Engineering, University of Pennsylvania, Philadelphia.

Pao, Y. H. (1978). An associate memory technique for the recognition of patterns, *Proc. 4th Int. Joint Conf. Pattern Recognition*, Kyoto, Japan, Nov. 7-10, 1978, pp. 405-407.

Pao, Y. H. (1981). A rule-based approach to electric power systems security assessment, *Proc. IEEE Conf. on Pattern Recognition and Image Processing*, Dallas, Tex., Aug. 1981, pp. 1-3.

Parikh, J. A., and Rosenfeld, A. (1978). Automatic segmentation and classification of infrared meteorological satellite data, *IEEE Trans. Syst. Man Cybern.*, vol. SMC-8, no. 10, pp. 736-743.

Patrick, E. A., and Shen, L. Y. L. (1971). Interactive use of problem knowledge for clustering and decision making, *IEEE Trans. Comput.*, vol. C-20, no. 2, pp. 216-222.

Patterson, J. D., Wagner, T. J., and Womack, B. F. (1967). A mean square performance criterion for adaptive pattern classification, *IEEE Trans. Autom. Control*, vol. 12, no. 2, pp. 195-197.

Pavel, M. (1979). Skeletons in pattern recognition categories, *Proc. IEEE Comput. Soc. Conf. Pattern Recognition Image Process.*, p. 406.

Pavlidis, T. (1978). Comments on "A new shape factor", *Comput. Graphics Image Process.*, vol. 8, no. 2, pp. 310-312.

Pavlidis, T. (1981). A flexible parallel thinning algorithm, *IEEE Comput. Soc. Conf. Pattern Recognition Image Process.*, Dallas, pp. 162-167.

Pavlidis, T., and Ali, F. (1979). A hierarchical syntactic shape analyzer, *IEEE Trans. Pattern Anal. Mach. Intell.*, vol. PAMI-1, no. 1, pp. 2-9.

Pavlidis, T., and Horowitz, S. L. (1974). Segmentation of plane curves, *IEEE Trans. Comput.*, vol. C-23, no. 8, pp. 860-870.

Perkins, W. A. (1982). A learning system that is useful for industrial inspection tasks, *Conf. Rec. 1982 Workshop Ind. Appl. Mach. Vision*, Research Triangle Park, N.C., May 3-5, 1982, pp. 160-167.

Persoon, E., and Fu, K. S. (1977). Shape discrimination using Fourier descriptors, *IEEE Trans. Syst. Man Cybern.*, vol. SMC-7, pp. 170-179.

Pfaltz, J. L., and Rosenfeld, A. (1969). Web grammars, *Proc. Joint Int. Conf. Artif. Intell.*, Washington, D.C.

Plott, H., Jr., Irwin, J., and Pinson, L. (1975). A real-time stereoscopic small computer graphics display system, *IEEE Trans. Syst. Man Cybern.*, vol. SMC-5, pp. 527-533.

Pollard, J. M. (1971). The fast Fourier transform in a finite field, *Math. Comput.*, vol. 25, no. 114, pp. 365-374.

Poppelbaum, W. J., Faiman, M., Casasent, D., and Sabd, D. S. (1968). On-line Fourier transform of video images, *Proc. IEEE*, vol. 56, no. 10, pp. 1744-1746.

Postaire, J. G., Vasseur, C. P. A. (1981). An approximate solution to normal mixture identification with application to unsupervised pattern classification, *IEEE Trans. on Pattern Analysis and Machine Intelligence*, vol. PAMI-3, no. 2, pp. 163-179.

Pratt, W. K. (1977). Pseudoinverse image restoration computational algorithms, in *Optical Information Processing* (G. W. Stroke, Y. Nesterikhin, and E. S. Barrekette, eds.), vol. 2, Plenum Press, New York, pp. 317-328.

Pratt, W. K., and Davarian, F. (1977). Fast computational techniques for pseudoinverse and Wiener image restoration, *IEEE Trans. Comput.*, vol. C-26, no. 6, pp. 571-580.

Pratt, W. K., and Kruger, R. P. (1972). Image processing over the ARPA computer network, *Proc. Int. Telemetering Conf.*, vol. 8, Los Angeles, Oct. 10-12, 1972, pp. 344-352.

Preston, K., Jr. (1971). Feature extraction by Goley hexagonal pattern transforms, *IEEE Trans. Comput.*, vol. C-20, pp. 1007-1014.

Preston, K. (1972). A comparison of analog and digital techniques for pattern recognition, *Proc. IEEE*, vol. 60, pp. 1216-1231.

Prewitt, J. M. S. (1970). Object enhancement and extraction, in *Picture Processing and Psychopictorics* (B. S. Lipkin and A. Rosenfeld, eds.), Academic Press, New York, pp. 75-149.

Price, C., Snyder, W., and Rajala, S. (1981). Computer tracking of moving objects using a Fourier domain filture based on a model of the human visual system, *IEEE Comput. Soc. Conf. Pattern Recognition Image Process.*, Dallas, Aug. 3-5, 1981.

Price, K. E. (1976). Change detection and analysis of multispectral images, Dept. of Computer Science, Carnegie-Mellon University, Pittsburgh, Pa.

Rahman, M. M., Jacquot, R. G., Quincy, E. A., and Stewart, E. A. (1980). Pattern recognition techniques in cloud research: II. Application, *Proc. 5th. Int. Conf. Pattern Recognition*, Miami Beach, Fla., Dec. 1-4, 1980 (IEEE, New York, 1980), pp. 470-474.

Rauch, H. E., and Firschein, O. (1980). Automatic track assembly for threshold infrared images, *Proc. Soc. Photo-opt. Instrum. Eng.*, vol. 253, pp. 75-85.

Ready, P. J., and Wintz, P. A. (1973). Information extraction, SNR improvement and data compression in multispectral imagery, *IEEE Trans. Commun.*, vol. COM-21, no. 10.

Reed, S. K. (1972). Pattern recognition and categorization, *Cognit. Psychol.*, vol. 3, pp. 382-407.

Reeves, A. P., and Rostampour, A. (1982). Computational cost of image registration with a parallel binary array processor, *IEEE Trans. Pattern Anal. Mach. Intell.*, vol. PAMI-4, no. 4, pp. 449-455.

Ridfway, W. C. (1962). An adaptive logic system with generalizing properties, *Stanford Electron. Lab. Tech. Rep. 1556-1*, Stanford University, Stanford, Calif.

Riseman, E. A., and Arbib, M. A. (1977). Computational techniques in visual systems: Part II: Segmenting static scenes, *IEEE Comput. Soc. Repository R77-87.*

Roach, J. W., and Aggarwal, J. K. (1979). Computer tracking of objects moving in space, *IEEE Trans. Pattern Anal. and Mach. Intell.*, vol. PAMI-1, no. 2, pp. 127-135.

Robbins, G. M. (1970). Image restoration for a class of linear spatially variant degradations, *Pattern Recognition*, vol. 2, no. 2, pp. 91-105.

Robbins, H., and Monro, S. (1951). A stochastic approximation method, *Ann. Math. Stat.*, vol. 22, pp. 400-407.

Robinson, G. S., and Frei, W. (1975). Final research report on computer processing of ERTS images, *USC-IPI Rep. 640*, Image Processing Institute, University of Southern California, Los Angeles.

Rogers, D., and Tanimoto, T. (1960). A computer program for classifying plants, *Science*, vol. 132, pp. 1115-1118.

Rosenblatt, F. (1957). The perceptron: a perceiving and recognizing automation, Project PARA, *Cornell Aeronaut. Lab. Rep. 85-460-1.*

Rosenblatt, F. (1960). On the convergence of reinforcement procedures in simple perceptrons, *Cornell Aeronaut. Lab. Rep. VG-1196-G4.*

Rosenfeld, A. (1969). Picture processing by computer, *Comput. Surv.*, vol. 1, no. 3, pp. 147-176.

Rosenfeld, A. (1974). Compact figures in digital pictures, *IEEE Trans. Syst. Man Cybern.*, vol. 4, pp. 211-213.

Rosenfeld, A. (1978). Clusters in digital picture, *Inf. Control*, vol. 39, no. 1, pp. 19-34.

Rosenfeld, A. (1978). Relaxation methods in image processing and analysis, *Proc. 4th Int. Joint Conf. Pattern Recognition*, Kyoto, Japan, Nov. 7-10, 1978, pp. 181-185.

Rosenfeld, A. (1982). Picture processing: 1981, *Comput. Graphics Image Process.*, vol. 19, no. 1, pp. 35-75, May 1982.

Rosenfeld, A., and Pfaltz, J. L. (1968). Distance functions on digital pictures, *Pattern Recognition*, vol. 1, pp. 33-61.

Rosenfeld, A., Fried, C., and Orton, J. N. (1965). Automatic cloud interpretation, *Photogramm. Eng.*, vol. 31, pp. 991-1002.

Rubin, L. M., and Frei, R. L. (1979). New approach to forward looking infrared (FLIR) segmentation, *Proc. Soc. Photo-opt. Instrum. Eng.*, vol. 205, pp. 117-125.

Ruell, H. E. (1982). Pattern recognition in data entry, *1982 Int. Zurich Semin. Digital Commun. Man-Mach. Interact.*, Zurich, Mar. 9-11, 1982 (IEEE, New York, 1982), pp. E2/129-136.

Salari, E., and Siy, P. (1982). A grey scale thinning algorithm using contextual information, *First Annu. Phoenix Conf. Comput. Commun.*, Phoenix, Ariz., May 9-12, 1982, pp. 36-38.

Sammon, J. W., Jr., Connell, D. B., and Opitz, B. K. (1971). Program for on-line pattern analysis, *Tech. Rep. TR-177* (2 vols.), Tome Air Development Center, Rome, Sept. 1971, AD-732235 and AD-732236.

Sanderson, A. C., and Segen, J. (1980). A pattern-directed approach to signal analysis, *Proc. 5th Int. Conf. Pattern Recognition*, Miami Beach, Fla., Dec. 1-4, 1980 (IEEE, New York, 1980).

Saridis, G. N. (1980). Pattern recognition and image processing theories, *Proc. 5th Int. Conf. Pattern Recognition*, Miami Beach, Fla., Dec. 1-4, 1980 (IEEE, New York, 1980).

Sawchuk, A. A. (1972). Space-variant image motion degradation and restoration, *Proc. IEEE*, vol. 60, pp. 854-861.

Scaltock, J. (1982). A survey of the literature of cluster analysis, *Comput. J.*, vol. 25, no. 1, pp. 130-133.

Schachter, B. (1978). A non-linear mapping algorithm for large data sets, *Comput. Graphics Image Process.*, vol. 8, no. 2, pp. 271-276.

Schell, R. R., Kodres, U. R., Amir, H., and Tao, T. F. (1980). Processing of infrared images by multiple microcomputer system, *Proc. Soc. Photo-opt. Instrum. Eng.*, vol. 241, pp. 267-278.

Schreiber, W. F. (1978). Image processing for quality improvement, *Proc. IEEE*, vol. 66, no. 12, pp. 1640-1651.

Sclove, S. L. (1981). Pattern recognition in image processing using interpixel correlation, *IEEE Trans. Pattern Anal. Mach. Intell.*, vol. PAMI-3, no. 2, pp. 206-208.

Sethi, I. K. (1981). A fast algorithm for recognizing nearest neighbors, *IEEE Trans. Syst. Man Cybern.*, vol. SMC-11, no. 3, pp. 245-248.

Shanmugar, K. S., and Paul, C. (1982). A fast edge thinning operator, *IEEE Trans. Syst. Man Cybern.*, vol. SMC-12, no. 4, pp. 567-569.

Shapiro, S. D. (1978). Properties of transforms for the detection of curves in noisy pictures, *Comput. Graphics Image Process.*, vol. 8, no. 2, pp. 219-236.

Shirai, Y., and Tsuji, S. (1972). Extraction of the line drawing of three dimensional objects by sequential illumination from several directions, *Pattern Recognition*, vol. 4, pp. 343-351.

Short, R. D., and Fukunaga, K. (1982). Feature extraction using problem localization, *IEEE Trans. Pattern Anal. Mach. Intell.*, vol. PAMI-4, no. 3, pp. 323-326.

Silberberg, T., Peleg, S., and Rosenfeld, A. (1981). Multiresolution pixel linking for image smoothing and segmentation, *Proc. SPIE Int. Soc. Opt. Eng.*, vol. 281, pp. 217-223.

Simon, J. C. (1978). Some current topics in clustering in relation with pattern recognition, *Proc. 4th Int. Joint Conf. Pattern Recognition*, Kyoto, Japan, Nov. 7-10, 1978, pp. 19-29.

Singleton, R. C. (1962). A test for linear separability as applied to self-organizing machines, in *Self-Organizing Systems* (M. C. Yovits, G. T. Jacobi, and G. D. Goldstein, eds.), Spartan Books, Washington, D.C.

Skinner, C. W., and Gonzalez, R. C. (1973). On the management and processing of earth resources information, *Proc. Conf. Mach. Process. Remotely Sensed Data*, Purdue University, Lafayette, Ind.

Sklansky, J. (1978). On the Hough technique for curve detection, *IEEE Trans. Comput.*, vol. C-27, no. 10, pp. 923-926.

Sklansky, J., Cordella, L. P., and Levialdi, S. (1976). Parallel detection of concavities in cellular blobs, *IEEE Trans. Comput.*, vol. C-25, no. 2, pp. 187-196.

Slepian, D. (1967). Restoration of photographs blurred by image motion, *Bell Syst. Tech. J.*, vol. 40, pp. 2353-2362.

Smith, S. P., and Jain, A. K. (1982). Structure of multi-dimensional patterns, *Proc. PRIP '82*, pp. 2-7.

Snyder, H. L. (1973). Image quality and observer performances, in *Perception of Displayed Information* (L. M. Biberman, ed.), Plenum Press, New York, pp. 87-118.

Solanki, J. K. (1978). Linear and nonlinear filtering for image enhancement, *Comput. Electron. Eng.*, vol. 5, no. 3, pp. 283-288.

Sondhi, M. M. (1972). Image restoration: the removal of spatially invariant degradations, *Proc. IEEE*, vol. 60, pp. 842-853.

Specht, D. F. (1967). Generation of polynomial discriminant functions for pattern recognition, *IEEE Trans. Electron. Comput.*, vol. EC-16, no. 3, pp. 308-319.

Spragins, J. (1966). Learning without a teacher, *IEEE Trans. Inf. Theory*, vol. IT-12, no. 2, pp. 223-230.

Srivastava, J. N. (1973). An information function approach to dimensionality analysis and curved manifold clustering, in *Multivariate Analysis*, vol. 3 (P. R. Krishnaiah, ed.), Academic Press, New York, pp. 369-382.

Starkov, M. A. (1981). Statistical model of images, *Avtometriya* (USSR), no. 6, pp. 6-12 (in Russian).

Stefanelli, R., and Rosenfeld, A. (1971). Some parallel thinning algorithms for digital pictures, *J. Assoc. Comput. Mach.*, vol. 18, pp. 255-264.

Stiefeld, B. (1975). An interactive graphics general purpose NDE (nondestructive evaluations) laboratory tool, *Proc. IEEE*, vol. 63, no. 10 (special issue on laboratory automation), pp. 1431-1437.

Stockham, T. G., Jr. (1972). Image processing in the context of a visual model, *Proc. IEEE*, vol. 60, no. 7, pp. 828-842.

Swain, P. H. (1970). On nonparametric and linguistic approaches to pattern recognition, Ph.D. dissertation, Purdue University, Lafayette, Ind.

Sze, T. W. (1979). Lecture note on digital image processing, Shanghai Jiao Tong University, Shanghai, China.

Takiyama, R. (1981). A committee machine with low committees, *IEEE Comput. Soc. Conf. Pattern Recognition Image Process.*, Dallas, Aug. 3-5, 1981.

Takiyama, R. (1982). A committee machine with a set of networks composed of two single-threshold elements as committee members, *Pattern Recognition*, vol. 15, no. 5, pp. 405-412.

Tanimoto, S., and Pavlidis, T. (1975). A hierarchical data structure for picture processing, *Comput. Graphics Image Process.*, vol. 4, pp. 104-119.

Tarter, M., and Silver, A. (1975). Implementation and application of bivariate Gaussian mixture decomposition, *J. Am. Stat. Assoc.*, vol. 70, no. 349, pp. 47-55.

Taylor, W. E., Jr. (1981). A general purpose image processing architecture for "real-time" and near "real-time" image exploitation,

IEEE SoutheastCon 1981 Congr. Proc., Huntsville, Ala., Apr. 5-8, 1981, pp. 646-649.

Tenenbaum, J. M., Barrow, H. G., Bolles, R. C., Fischler, M. A., and Wolf, H. C. (1979). Map-guided interpretation of remotely sensed imagery, *Proc. 1979 IEEE Comput. Soc. Conf. Pattern Recognition Image Process.*, pp. 610-617.

Thatcher, J. W. (1973). Tree automata: an informal survey, in *Current Topics on the Theory of Computing* (A. V. Aho, ed.), Prentice-Hall, Englewood Cliffs, N.J.

Thomas, J. C. (1971). Phasor diagrams simplify Fourier transforms, *Electron. Eng.*, pp. 54-57.

Thomason, M. G., Barrero, A., and Gonzalez, R. C. (1978). Relational database table representation of scenes, *Proc. IEEE SoutheastCon*, Atlanta, pp. 32-37.

Tomita, F., Yachida, M., and Tsuji, S. (1973). Detection of homogeneous regions by structural analysis, *Proc. Int. Joint Conf. Artif. Intell.*, Stanford, Calif., Aug. 1973, pp. 564-571.

Tomoto, Y., and Taguchi, H. (1978). A new type image analyzer and its algorithm, *1978 Int. Congr. Photogr. Sci.*, Rochester, N.Y., Aug. 1978, pp. 20-26. Also SPSE, 1978, pp. 229-231.

Toney, E. (1983). Image enhancement through clustering specification, Masters Thesis, Pennsylvania State University, University Park, May 1983.

Tou, J. T. (1968). Feature extraction in pattern recognition, *Pattern Recognition*, vol. 1, no. 1, pp. 2-11.

Tou, J. T. (1968). Information theoretical approach to pattern recognition, *IEEE Int. Conv. Rec.*

Tou, J. T. (1969). Engineering principles of pattern recognition, in *Advances in Information Systems Science* (J. T. Tou, ed.), vol. 1, Plenum Press, New York.

Tou, J. T. (1969). Feature selection for pattern recognition system, in *Methodologies of Pattern Recognition* (S. Watanabe, ed.), Academic Press, New York.

Tou, J. T. (1969). On feature encoding in picture processing by computer, *Proc. Allerton Conf. Circuits Syst. Theory*, University of Illinois, Urbana.

Tou, J. T. (1972). Automatic analysis of blood smear micrographs, *Proc. 1972 Comput. Image Process. Recognition Symp.*, University of Missouri, Columbia.

Tou, J. T. (1972). CPA: a cellar picture analyzer, *IEEE Comput. Soc. Workshop Pattern Recognition*, Hot Springs, Va.

Tou, J. T. (1979). DYNOC — a dynamic optimal cluster-seeking technique, *Int. J. Comput. Inf. Sci.*, vol. 8, no. 6, pp. 541-547.

Tou, J. T., and Gonzalez, R. C. (1971). A new approach to automatic recognition of handwritten characters, *Proc. Two-Dimensional Signal Process. Conf.*, University of Missouri, Columbia.

Tou, J. T., and Gonzalez, R. C. (1972). Automatic recognition of handwritten characters via feature extraction and multilevel decision, *Int. J. Comput. Inf. Sci.*, vol. 1, no. 1, pp. 43-65.

Tou, J. T., and Gonzalez, R. C. (1972). Recognition of handwritten characters by topological feature extraction and multilevel categorization, *IEEE Trans. Comput.*, vol. C-21, no. 7, pp. 776-785.

Tou, J. T., and Heydorn, R. P. (1967). Some approaches to optimum feature extraction, in *Computer and Information Science*, vol. II (J. T. Tou, ed.), Academic Press, New York.

Triendle, E. E. (1971). An image processing and pattern recognition system for time variant images using TV cameras and a matrix computer, *Artif. Intell.*, *AGARD Conf. Proc.*, London, 1971, Paper 23.

Triendle, E. E. (1979). Landsat image processing, *Advances in Digital Image Processing*, Plenum, New York, 1979, pp. 165-175.

Tsao, Y. F., and Fu, K. S. (1981). Parallel thinning operations for digital binary images, *IEEE Comput. Soc. Conf. Pattern Recognition Image Process.*, Dallas, Aug. 3-5, 1981, pp. 150-155.

Tsypkin, Ya. Z. (1965). Establishing characteristics of a function transformer from randomly observed points, *Autom. Remote Control*, vol. 26, no. 11, pp. 1878-1882.

Twogood, R. E., and Ekstrom, M. P. (1976). An extension of Eklundh's matrix transposition algorithm and its applications in digital image processing, *IEEE Trans. Comput.*, vol. C-25, no. 9, pp. 950-952.

Uhr, L. (1971). Flexible linguistic pattern recognition, *Pattern Recognition*, vol. 3, no. 4, pp. 363-383.

Uhr, L. (1971). Layered recognition cone networks that preprocess, classify and describe, *Proc. Two-Dimensional Signal Process. Conf.*, Columbia, Missouri, pp. 311-312.

Urquhart, R. (1982). Graph theoretical clustering based on limited neighborhood sets, *Pattern Recognition*, vol. 15, no. 3, pp. 173-187.

VanderBrug, G. J., and Nagel, R. N. (1979). Vision systems for manufacturing, *Proc. 1979 Joint Autom. Control Conf.*, Denver, June 17-21, 1979, pp. 760-770.

VanderBrug, G. J., and Rosenfeld, A. (1977). Two-stage template matching, *IEEE Trans. Comput.*, vol. 26, no. 4, pp. 384-394.

VanderBrug, G. J., and Rosenfeld, A. (1978). Linear feature mapping, *IEEE Trans. Syst. Man Cybern.*, SMC-8, no. 10, pp. 768-774.

Vickers, A. L., and Modestino, J. W. (1981). A maximum likelihood approach to texture classification, *IEEE Trans. Pattern Anal. Mach. Intell.*, vol. PAMI-4, no. 1, pp. 61-68.

Vilione, S. S. (1970). Applications of pattern recognition technology, in *Adaptive Learning and Pattern Recognition Systems: Theory and Applications* (J. M. Mendal and K. S. Fu, eds.), Academic Press, New York, pp. 115-162.

Vinea, A., and Vinea, V. (1971). A distance criterion for figural pattern recognition, *IEEE Trans. Comput.*, vol. C-20, June 1971, pp. 680-685.

Wald, L. H., Chou, F. M., and Hines, D. C. (1980). Recent progress in extraction of targets out of cluster, *Proc. Soc. Photoopt. Instrum. Eng.*, vol. 253, pp. 40-55.

Walter, C. M. (1968). Interactive systems applied to the reduction and interpretation of sensor data, *Proc. Digital Equipment Comput. Users Fall Symp.*, Dec. 1968.

Walter, C. M. (1969). Comments on interactive systems applied to the reduction and interpretation of sensor data, *IEEE Comput. Commun. Conf. Rec.*, 1969 (IEEE Spec. Publ. 69, 67-MVSEC), pp. 109-112.

Wang, C. Y., and Wang, P. P. (1982). Pattern analysis and recognition based upon the theory of fuzzy subsets, *Conf. Proc. IEEE SoutheastCon '82*, Destin, Fla., Apr. 4-7, 1982, pp. 353-355.

Wang, S., Rosenfeld, A., and Wu, A. Y. (1982). A medial axis transformation for grey scale pictures, *IEEE Trans. Pattern Anal. Mach. Intell.*, vol. PAMI-4, no. 4, pp. 419-421.

Warmack, R. E., and Gonzalez, R. C. (1972). Maximum error pattern recognition in supervised learning environments, *IEEE Conv. Rec. — Region III*.

Warmack, R. E., and Gonzalez, R. C. (1973). An algorithm for the optimal solution of linear inequalities and its application to pattern recognition, *IEEE Trans. Comput.*, vol. C-22, pp. 1065-1075.

Watanabe, S. (1965). Karhunen-Loeve expansion and factor analysis theoretical remarks and applications, *Proc. 4th Conf. Inf. Theory*, Prague.

Watanabe, S. (1970). Feature compression, in *Advances in Information Systems Science* (J. T. Tou, ed.), vol. 3, Plenum Press, New York.

Watanabe, W. (1971). Ungrammatical grammar in pattern recognition, *Pattern Recognition*, vol. 3, no. 4, pp. 385-408.

Wechsler, H. (1979). Feature extraction for texture discrimination, *Proc. 1979 IEEE Comput. Soc. Conf. Pattern Recognition Image Process.*, Chicago, pp. 399-403.

Weszka, J. S., and Rosenfeld, A. (1979). Histogram modification for threshold selection, *IEEE Trans. Syst. Man Cybern.*, vol. SMC-9, no. 1, pp. 38-52.

Whitmore, P. G., Rankin, W. C., Baldwin, R. D., and Garcia, A. (1972). Studies of aircraft recognition training, *Tech. Rep.*, Human Research Organization, Alexandria, Va., AD-739923.

Whitney, A. W., and Blasdell, W. E. (1971). Signal analysis and classification by interactive computer graphics, in *AGARD, Artificial Intelligence*, General Electric Co., Syracuse, N.Y., 1971.

Widrow, B. (1962). Generalization and information storage in net-
works of Adaline neurons, in *Self-Organizing Systems* (M. C.
Yovits, G. T. Jacobi, and D. Goldstein, eds.), Spartan Books,
Washington, D.C.

Widrow, B. (1973). The rubber mask technique: I. Pattern measure
and analysis, *Pattern Recognition*, vol. 5, no. 3, pp. 175-197.

Widrow, B. (1973). The rubber mask technique: II. Pattern storage
and recognition, *Pattern Recognition*, vol. 5, no. 3, pp. 199-211.

Will, P. M., and Koppleman, G. M. (1971). MFIPS: a multi-functional
digital image processing system, *IBM Res.*, *RC 3313*, Yorktown
Heights, N.Y., 1971.

Will, P. M., and Pennington, K. S. (1972). Grid coding: a novel
technique for image processing, *Proc. IEEE*, vol. 60, no. 6,
pp. 669-680.

Winder, R. O. (1963). Bounds on threshold gate realizability, *IEEE
Electron. Comput.*, vol. EC-12, no. 4, pp. 561-564.

Winder, R. O. (1968). Fundamentals of threshold logic, in *Applied
Automata Theory* (J. T. Tou, ed.), Academic Press, New York.

Wolfe, J. H. (1970). Pattern clustering by multivariate mixture analysis, *Multivariate Behav. Res.*, vol. 5, p. 329.

Wolferts, K. (1974). Special problems in interacting image processing
for traffic analysis, *Proc. 2nd Int. Joint Conf. Pattern Recognition*, Copenhagen, Aug. 13-15, 1974, pp. 1-2.

Wong, M. A., and Lane, T. (1982). A kth nearest neighbor clustering procedure, *Computer Science and Statistics: Proc. 13th
Symp. Interface*, Pittsburgh, Pa., Mar. 12-13, 1981, pp. 308-
311.

Wong, R. Y. (1977). Image sensor transformations, *IEEE Trans. Syst.
Man Cybern.*, vol. SMC-7, no. 12, pp. 836-841.

Wong, R. Y. (1982). Pattern recognition with multi-microprocessors,
Proc. IEEE 1982 Region 6 Conf., Anaheim, Calif., Feb. 16-19,
1982, pp. 125-129.

Wong, R. Y., and Hall, E. L. (1977). Sequential hierarchical scene
matching, *IEEE Trans. Comput.*, vol. C-27, no. 4, pp. 359-365.

Wong, R. Y., Lee, M. L., and Hardaker, P. R. (1980). Airborne
video image enhancement, *Proc. Soc. Photo-opt. Instrum. Eng.*,
vol. 241, pp. 47-50.

Wood, R. E., and Gonzalez, R. C. (1981). Real time digital enhancement, *Proc. IEEE*, vol. 69, no. 5, pp. 643-654.

Wu, C. L. (1980). Considerations on real time processing of space-
borne aperture radar data, *Proc. Soc. Photo-opt. Instrum. Eng.*,
vol. 241, pp. 11-19.

Wu, S. Y., Dubitzki, T., and Rosenfeld, A. (1981). Parallel computation of contour properties, *IEEE Trans. Pattern Anal. Mach.
Intell.*, vol. PAMI-3, no. 3, pp. 331-337.

Yokoyama, R., and Haralick, R. M. (1978). A texture pattern synthesis method by growth method, *Tech. Rep.*, Iwata University, Morioka, Japan.

Young, I. T. (1978). Further consideration of sample and feature size, *IEEE Trans. Inf. Theory*, vol. IT-24, no. 6, pp. 773-775.

Zahn, C. T. (1971). Graph theoretical methods for detecting and describing gestalt clusters, *IEEE Trans. Comput.*, vol. C-20, no. 1, pp. 68-86.

Zweig, H. J., Barrett, E. B., and Hu, P. C. (1975). Noise cheating image enhancement, *J. Opt. Soc. Am.*, vol. 65, no. 11, pp. 1347-1353.

Index